A CULTURAL HISTORY OF PLANTS

VOLUME 4

A Cultural History of Plants
General Editors: Annette Giesecke and David J. Mabberley

Volume 1
A Cultural History of Plants in Antiquity
Edited by Annette Giesecke

Volume 2
A Cultural History of Plants in the Post-Classical Era
Edited by Alain Touwaide

Volume 3
A Cultural History of Plants in the Early Modern Era
Edited by Andrew Dalby and Annette Giesecke

Volume 4
A Cultural History of Plants in the Seventeenth and Eighteenth Centuries
Edited by Jennifer Milam

Volume 5
A Cultural History of Plants in the Nineteenth Century
Edited by David J. Mabberley

Volume 6
A Cultural History of Plants in the Modern Era
Edited by Stephen Forbes

A CULTURAL HISTORY OF PLANTS

IN THE SEVENTEENTH AND EIGHTEENTH CENTURIES

VOLUME 4

Edited by Jennifer Milam

BLOOMSBURY ACADEMIC
LONDON • NEW YORK • OXFORD • NEW DELHI • SYDNEY

BLOOMSBURY ACADEMIC
Bloomsbury Publishing Plc
50 Bedford Square, London, WC1B 3DP, UK
1385 Broadway, New York, NY 10018, USA
29 Earlsfort Terrace, Dublin 2, Ireland

BLOOMSBURY, BLOOMSBURY ACADEMIC and the Diana logo
are trademarks of Bloomsbury Publishing Plc

First published in Great Britain 2022
Paperback edition published in 2025

Series design: Raven Design
Cover image © Archives Charmet/Bridgeman Images

A catalogue record for this book is available from the British Library.

Library of Congress Control Number: 2021932844

ISBN: PB: 978-1-3505-5061-2
Set: 978-1-3505-5064-3
ePUB: 978-1-3502-5933-1
ePDF: 978-1-3502-5934-8

Series: The Cultural History Series

Typeset by Integra Software Services Pvt. Ltd.
Printed and bound in Great Britain

CONTENTS

ILLUSTRATIONS

SERIES PREFACE

The connectedness of humans to plants is the most fundamental of human relationships. Plants are, and historically have been, sources of food, shelter, bedding, tools, medicine, and, most importantly, the very air we breathe. Plants have inspired awe, a sense of well-being, religious fervor, and acquisitiveness alike. They have been collected, propagated, and mutated, as well as endangered or driven into extinction by human impacts such as global warming, deforestation, fire suppression, and over-grazing. *A Cultural History of Plants* traces the global dependence of human life and civilization on plants from antiquity to the twenty-first century and comprises contributions by experts and scholars in a wide range of fields, including anthropology, archaeology, art history, botany, classics, garden history, history, literature, and environmental studies more broadly. The series consists of six illustrated volumes, each devoted to an examination of plants as grounded in, and shaping, the cultural experiences of a particular historical period. Each of the six volumes, in turn, is structured in the same way, beginning with an introductory chapter that offers a sweeping view of the cultural history of plants in the period in question, followed by chapters on plants as staple foods, plants as luxury foods, trade and exploration, plant technology and science, plants and medicine, plants in (popular) culture, plants as natural ornaments, and the representation of plants. This cohesive structure offers readers the opportunity both to explore a meaningful cross-section of humans' uses of plants in a given period and to trace a particular use—as in medicine, for example—through time from volume to volume. The six volumes comprising *A Cultural History of Plants* are as follows:

Volume 1: *A Cultural History of Plants in Antiquity* (*c*. 10,000 BCE–500 CE)
Volume 2: *A Cultural History of Plants in the Post-Classical Era* (500–1400)
Volume 3: *A Cultural History of Plants in the Early Modern Era* (1400–1650)
Volume 4: *A Cultural History of Plants in the Seventeenth and Eighteenth Centuries* (1650–1800)
Volume 5: *A Cultural History of Plants in the Nineteenth Century* (1800–1920)
Volume 6: *A Cultural History of Plants in the Modern Era* (1920–present).

By way of guidance to our readers, it should be noted that the plant names used in these volumes accord with those in the fourth edition of *Mabberley's Plant-book* (Cambridge University Press, 2017). When they are discussed, individual plants are identified using their common names and, at their first mention in each chapter, with their scientific names: e.g. bay laurel (*Laurus nobilis*). As is recommended for general works such as this, the authorities to whom the scientific names are attributed (e.g. *Laurus nobilis* L., where L. identifies Linnaeus as the identifying authority) have been omitted.

Annette Giesecke and David J. Mabberley,
General Editors

Introduction

Plants and Culture during the Enlightenment

JENNIFER MILAM AND GARRITT VAN DYK

This volume, the fourth in the series, considers the role of plants as they intersect with imperial power and empirical philosophy during the Enlightenment. While the Enlightenment was not a global movement, the ideas that developed during this period impacted many parts of the world through exploration, trade, and colonization. Framed by the beginning of the seventeenth century with the rise of absolute monarchies in Europe and the decades of revolution at the end of the eighteenth century, the Enlightenment was a transformative period in history that established new norms of understanding and behavior. Outside of Europe, powerful dynasties in the Ottoman Empire, Mughal India, and Imperial China continued to rule large portions of the globe. The intensification of international commerce demanded diplomatic relationships and promoted cultural traffic between Europe and Asia. Voyages of exploration expanded European knowledge of North America and the southern hemisphere. Commerce and exploration increased the opportunities for contact and exchange. Ultimately, the circulation of goods and ideas made the world a much smaller place.

Greater contact with other cultures through trade and diplomacy led to the circulation of plant specimens from around the globe. Botany in the seventeenth and eighteenth centuries was both a theoretical and practical discipline, studied by academics at state-sponsored institutions such as the botanical gardens of Leiden and Padua. These institutions worked with the joint-stock companies from England, the Netherlands, and France to collect as many new species of plants as possible. Once in Europe, the challenge to organize and classify the unknown plants began. The potential usage of the plants and experimentation with their acclimatization and propagation were additional hurdles to incorporating the exotic species into the cultural fabric of everyday life. Indigenous knowledge of the plants as food, medicine, shelter, dyes, and pigments was used as a starting point for the exploitation of new species, which often required mediation or translation to adapt to European and American cultural norms. This process was based on empiricism, and resulted in a vast expansion of knowledge and systems of classification that enabled individuals to gain a better understanding of the natural world and their place within it. Networks of collectors, scientists, and philosophers engaged in an exchange of ideas based on an empirical understanding of nature that developed new ideas, new science, and fostered the development of new economies focused on global trade.

These discoveries took place during a period when there were significant changes in thinking: a move away from absolutism and religious authority toward liberal democracy; a shift from superstition to evidence-based rationality and empiricism; a new appreciation of other cultures through increased tolerance and cosmopolitanism; and a belief in individual responsibility and the possibility of contributing to making the world a better place. This overarching contextualization of the cultural history of plants between 1650–1800 spans a time when absolute monarchs reigned over their subjects with the authority of divine right in Europe, and powerful emperors ruled in China, India, and Turkey. It incorporates the project of the Enlightenment which developed scientific methods and theories of rationality that were embodied in the works of botanists and philosophers, and concludes in the United States, a product of the Enlightenment project, with the experimental gardens of the American statesman Thomas Jefferson (1743–1826) at Monticello, located just outside Charlottesville, Virginia. Jefferson's vision of an agrarian democracy and economic self-sufficiency punctuates the end of the period, before the onset of increased industrialization, Romanticism, and modernity.

Our introduction begins with a study of palaces as way of exploring how monarchs used the cultural language of plants to convey the divine right and grandeur of the ruler. We follow this with an analysis of how the pursuit of botanical knowledge enabled individuals outside the realm of royalty to classify a new universe of plants encountered through exploration and test the boundaries of how the natural world could be manipulated to allow plants to be cultivated outside of their natural environments. This interest in transplanting simultaneously contemplated the creation of an earthly Eden and presented a challenge to the Providential distribution of plants in different climes. Exotic plants were also manipulated to produce new hybrids or variations that had never been seen before.

Experimentation, using the scientific method and empiricism that drove the advancements in knowledge of the plant world, allowed for a better understanding of the place of humankind in the natural world. Beyond botanical practice, Enlightenment writers engaged with plants through discourse, employing metaphors of cultivation and metamorphosis to communicate philosophical concepts and political ideologies. These writers were not armchair botanists, however, and their hands-on practice provided an empirical basis for their philosophical and political thought. In practice, and in discourse, plants in the seventeenth and eighteenth century were deeply embedded in the cultural landscape through a relationship of mutual influence that guided, and was guided by, the major intellectual movements of the period.

PLANTS AND THE DISPLAY OF ABSOLUTE POWER

Before the Enlightenment ideals of liberal democracy developed, absolutist rulers conveyed their authority symbolically through control over nature in gardens and spaces dedicated to the cultivation and display of plants. While many elites had gardens and cultivated plants, the ultimate expression of prestige and power through display was the province of the monarch, who ruled by divine right. Despite this proscription, the Superintendent of Finance to the French king Louis XIV (r. 1643–1715), Nicolas Fouquet (1615–80), created an estate in 1661 with gardens and a château that eclipsed the grandeur of the Sun King. Over two decades Fouquet had transformed the grounds and château of Vaux-le-Vicomte in Maincy, France, into a harmonious masterpiece through the artistry of: Louis

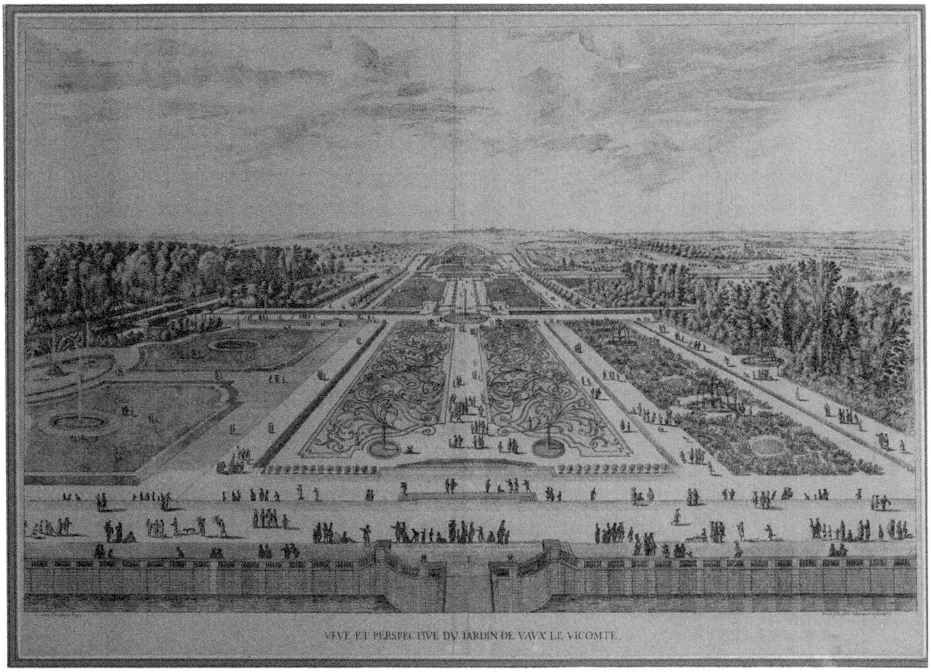

FIGURE 0.1 Israel Silvestre, *Vue et perspective du jardin de Vaux le Vicomte*, engraving, Château de Vaux le Vicomte. © Alamy.

Le Vau (1612–70), architect; Charles Le Brun (1619–90), painter and interior designer; and André Le Nôtre (1613–1700), the most eminent landscape gardener in France.

The gardens of Vaux-le-Vicomte provide a microcosm of the European perspective on the natural world at this point in history. It was something to be ordered and controlled. The central avenue extends from the axis of the château to the horizon, with symmetrical embroidered parterres on either side. Inside these beds, plants were used to make flowing ribbons of formalized foliate designs. Lying to one side were flowerbeds that were brought into geometric conformity, but still allowed to bloom. Fouquet had been an avid plant collector prior to building the gardens at Vaux-le-Vicomte. His plant collections cultivated at Saint-Mandé, just east of Paris, included rare shrubs and flowers, with some two hundred orange trees. The scale of the estate was unlike anything else in France, including any of Louis XIV's royal residences. Falsely accused, then convicted of embezzlement, Fouquet was disgraced and imprisoned, with his property seized by the King. Surrounded by jealous rivals, he was already a target before he invited Louis XIV to visit the most luxurious estate in France. Fouquet's family motto (*Quo non ascendet*—What heights will he not climb?) reflected his political and social ambitions, but incurred the wrath of the King when he usurped the symbolic role of the monarch as emperor of the natural world.

After his property was confiscated, Fouquet's vast collections of plants acted as a sort of nursery for Louis XIV's Versailles, which underwent a transformation to become the center of court life. Moved into the gardens of Versailles, the plants were ordered into a disciplined geometric scheme of *allées* through which the King and his court would make tours. Orange trees were displayed in tubs and arranged in straight formations, while new espalier patterns were developed for a variety of trees in the Potager du Roi, the royal fruit and vegetable gardens. Covering eight hectares with twenty-nine separate walled enclosures, all kinds of fruit were cultivated including pears and peach, and the exotic sago palm (*Cycas revoluta*). The royal gardener in charge of the Potager du Roi, Jean de La Quintinie (1626–88) completed his *Instructions pour les jardins fruitiers*, which was published posthumously, in 1690 and it became a classic reference for fruit pruning. Further plantings of ordered perfection at Versailles included the flowerbeds that surrounded Trianon, a pavilion destination within the gardens where the King would dine with ladies of the court. During the meal, the flowerbeds would be completely replanted with new arrangements and colors to delight guests with this vivid demonstration of the King's power over nature.

Versailles is the physical, spatial, and natural manifestation of the divine right of kings as the underlying principle of absolutism. The ordering and control of the gardens, as well as the cultivation, shaping, and display of plants, both domestic and exotic, aimed to demonstrate in parallel that as God gave order to the universe, so the King brought order to the realm. Effectively, the divine right of Louis XIV to rule as an absolute monarch was planted into the gardens of Versailles.

ALL UNDER THE HEAVEN IN IMPERIAL CHINA

Outside of Europe, the imperial dynasties of China, the Ottoman Empire, and Mughal India also expressed their authority and control of territory and nature through formal plantings attached to their palaces. In imperial China, the legitimacy of the emperor to rule was justified by the ancient concept of the Mandate of Heaven. According to this religious and political teaching, Heaven embodies the natural order of the universe and bestows the right to rule on a "Son of Heaven," the sacred imperial title of the Chinese emperor. The Garden of Perfect Brightness (Yuanming Yuan) planted this ideal into the natural environment. Begun by the Qing emperor Kangxi (r. 1661–1722), the gardens in the western suburbs of Beijing were brought to full glory by his grandson, Qianlong (r. 1736–96). It represented *Tianxia*, or All Under the Heaven, through the ordering of its various components that combined the gracefulness of country life with the grandeur of imperial power (Cha-Tsu Siu 2013: xxii–xxiv; Wong 2001: 20). Architectural elements harmonized with the distribution of *yin* (water) and *yang* (rocks or hills), while vegetation included willow trees (*Salix* spp.), bamboo groves, tall pines (*Pinus* spp.), water chestnuts (*Eleocharis dulcis*), lotus flowers (*Nelumbo nucifera*), chrysanthemums (*Chrysanthemum* x *morifolium*), orchids, juniper (*Juniperus communis*), and tree peonies (*Paeonia* x *suffruticosa*).

The disposition of plants in Chinese gardens during this period shared commonalities with Chinese landscape painting and poetry. In all three art forms, plants were represented and arranged through an attempt to emulate nature and demonstrate a genuine harmony between man and the natural world. Plants and trees had metaphoric value as well (Wong 2001: 9–10). At Yuanming Yuan vegetation carried symbolic associations that dated back to the Han Dynasty (202 BCE–220 CE): lotus flowers expressed harmony; juniper conveyed the theme of immortality; tree peonies were associated with royalty. In paintings of flowers created by Western Jesuit artists who worked in the Qing court, plants are arranged to symbolize sagacious rule.

FIGURE 0.2 Giuseppe Castiglione, *Gathering of Auspicious Signs*. Hanging scroll, ink and colors on silk, 173 × 86.1 cm, 1723. National Palace Museum, Taipei. © Alamy.

In 1792, the senior British diplomat George Macartney (1737–1806) was sent to China for the expansion of trade under the auspices of honoring Qianlong's eightieth birthday. Lodged in the neighborhood of Yuanming Yuan, members of the Macartney embassy recounted the "vast variety of elegant little buildings" glimpsed in the imperial gardens and the disposition of parts, "broken into hill and dale, and diversified with wood and lawn." The entire garden, it appeared to the British, had been "thrown up with immense labour in an irregular and, as it were, fortuitous manner, so as to represent the free hand of nature" (Wong 2001: 84). The embassy included two botanical gardeners who had been charged with gathering as many plants as possible, especially tea (Kitson 2013: 135–42). Joseph Banks (1743–1820), a prominent British botanist and president of the Royal Society, provided a list of Chinese plants that he was keen to source and requested that special techniques for "accelerating the blossoming of plants" and dwarfing trees be observed. Hundreds of plant specimens were collected on the Macartney embassy, with Banks imagining how pleasant it would be to "Learn that our King at Kew & and Emperor of China at Jehol solace themselves under the shade of many of the same trees & admire the elegance of many of the same flowers in their respective gardens" (Kitson 2013: 139). Such a poetic description of diplomacy formed through a botanical interconnectedness between cultures encapsulates the cosmopolitan ideals of the late eighteenth century.

CROSS-CULTURAL EXCHANGE IN MUGHAL INDIA

A richness of visual and textual sources from around the world demonstrates the value and significance of plants to a diverse range of cultures in the seventeenth and eighteenth centuries. They also record the circulation of goods and ideas that evidence the cultural exchange between East and West, driven by curiosity and commerce. Flowers, in particular, were a source of fascination that resulted in paintings and prints depicting individual plants as botanical specimens, decorative arrangements, and garden plantings. From Mughal India, the album of Prince Dara Shikoh (1615–59) contains several studies of flowers depicted in vibrant colors with stylized golden clouds floating above. The naturalness of the flowers, which allows many plants to be identified—such as pink roses, blue irises, pimpernels (*Anagallis*), and Jacob's Ladder (*Polemonium caeruleum*)—suggest the close observation of nature; yet it is most likely that the anonymous local artists were drawing on Dutch engravings. Visually, they combine elements of European botanical studies with Indian tastes for stylized relief designs. Flattened on the page, the individual flower studies demonstrate botanical accuracy and deliberate arrangement, evidencing their debt to Dutch flower engravings which circulated as florilegia at the time. Created in the early 1630s, this collection of paintings in folios interspersed with calligraphy was a gift for the Prince's bride, dedicated with an inscription to "his special companion, intimate and confidante" (Losty and Roy 2013). Around this time, flowers and floral arrangements began to dominate Mughal textiles, architecture, and album pages as the most desirable decorative motifs. While demonstrating a close interest in observing and recording the natural world, Mughal artists also used European prints as inspirations for their paintings and the decorative borders they created for imperial albums (Skelton 1972).

Trees, shrubs, and flowers are depicted in miniatures that record Mughal gardens, such as those composed at the Taj Mahal (1632–53), built by Shah Jahan (1592–1666) as a tomb for his wife. Trees which dominate the Indian landscape are often shown in these paintings, including banyans (*Ficus benghalensis*), palms (*Arecaceae*), mango, and plantains (types of banana), as are pink-flowered Indian lotus, irises, day-lilies (*Hemerocallis* spp.), poppies (*Papaver*), and oleanders (*Nerium oleander*). As the gardens were designed as a memorial, the meanings generated by the arrangement of plants conveyed messages about life and death. Elevated on a terrace, the tomb floats above the paradise garden, with its traditional arrangement into a quadripartite canal system that represents the four rivers

FIGURE 0.3 Anonymous, *A pink rose, a blue iris, a pimpernel and other flowering plants; gold clouds at top. Dara Shikoh Album,* 1630–40. Opaque watercolor. Gouache with gold; background uncolored. British Library, Add. Or. 3129, f. 67v. © Alamy.

FIGURE 0.4 The Taj Mahal. 'Amal-i Salih. India, eighteenth century. The Taj Mahal with European sightseers on the terrace. Angels pouring gold from the clouds. (Inscriptions on the building.) A miniature painting from an eighteenth-century manuscript of 'Amal-i Salih, a history of Shah Jahan. Image taken from 'Amal-i Salih. Originally published/produced in India, eighteenth century. British Library, Or. 2157, f. 612. © Alamy.

of life. An abundance of flowers decorated sunken beds, while fruit trees lined the main pools, to celebrate life and fertility. In contrast, cypress avenues lined the main pool and symbolized death (Hobhouse 1992: 66–7).

ATTITUDES TOWARDS NATURE IN THE OTTOMAN EMPIRE

In contrast to the well-documented plantings in China and Mughal India, there are few records of Turkish gardens from the Ottoman period. The landscape murals that decorate the interior of the Queen Mother's apartments in the harem at Topkapi Palace provide idealized images of the largely natural plantings, as well as Turkish attitudes towards nature. Placed in the upper register of the walls, the paintings record landscapes that unfold along the curving banks of rivers. A detail of one section depicts a delicate kiosk and rectangular pool, surrounded by geometric flowerbeds punctuated by towering cypress trees (*Cupressus*) shaped into cones and intertwined with flowering almonds (*Prunus dulcis*). Yet wilderness is visibly present just beyond the garden walls. While plantings could be shaped and controlled, the implication of this contrast is that the Turks preferred uncultivated nature to be always near. Similarly, the decorative ornamentation that frames the murals are derived from natural forms. The dome is a forest of branches and vines that breaks apart the architectural order to put culture into tension with nature.

A CULTURE OF FLOWERS

In the seventeenth century, the increased circulation of goods and people brought Europeans into contact more frequently with both the Moghul and Ottoman Empires as they traded in goods sourced from overland Asian trade routes. Early Modern globalization led to cultural interactions that introduced European travelers, merchants, and diplomats to new plants and comestibles from Central Asia and Eastern Europe. Beyond the world of commerce, travelers relayed tales of fields filled with cultivated flowers. At the end of the sixteenth century, the Flemish herbalist and diplomat, Ogier Ghiselin de Busbecq (1522–92) in the employ of the Austrians as ambassador to Süleyman the Magnificent, wrote about the beauty of the unknown flowers he had encountered: "on the last stage of our journey to Constantinople ... we everywhere came across quantities of flowers—narcissi, hyacinths, and tulipans, as the Turks call them ... they possess so wonderful a scent that a large quantity of them cause a headache ... The tulip has no scent but is admired for its beauty and the variety of its colours. The Turks are very fond of flowers" (Forster 1927: 24). As early as 1630, the Ottomon explorer Evliya Çelebi (1611–82) described over three hundred florists and eighty flower shops in Adrianople (present-day Edirne). In Constantinople (present-day Istanbul), flowers were found for sale in the marketplace, in gardens, and as a ubiquitous element in decorative arts. The ornamental gardens of Constantinople and the palace gardens were filled with jonquils (*Narcissus jonquilla*), hyacinths (*Hyacinthus orientalis*), irises, anemones, buttercups (*Ranunculus asiaticus*), and tulips (Baytop 1993: 51). Outside the garden, flowers were prized as a decorative motif and appeared in architecture, textiles, illuminated manuscripts, paintings, and tiles (Demiriz 1993: 58).

Of all these flowers, tulips made the biggest impression on European visitors and, beginning in the sixteenth century, were imported from the Ottoman Empire. By the

FIGURE 0.5 Mural paintings inside the apartments of the Queen Mother in the Harem of Topkapi Palace, Instanbul, Turkey. © Alamy.

seventeenth century, tulips had become an obsession in the Netherlands, with this "tulipmania" leading to financial speculation through a rudimentary form of a futures market. Britain did not experience the boom and bust of the tulip bubble but was nonetheless influenced by the new Turkish flower. The English botanist John Parkinson's (1567–1660) *Paradisi in Sole Paradisus Terrestris* listed more than 100 cultivars of tulip which he admired above all others for their beauty and variety: "this flower ... deserveth his true commendations and acceptance with all lovers of these beauties, both for the stately aspect, and for the admirable variety of the colours that doe arise in them" (Parkinson 1629: 45). The variegation Parkinson saw was the product of a viral disease affecting or breaking the flower's pigmentation, producing streaks and speckles. He was unaware of the virus but observed that weaker plants changed from solid to striped. Parkinson also noted that attempting to cultivate the identical flower from seed was unpredictable, "yielding a mixture and variety that hath not before been observed" (Parkinson 1629: 45). The petal pattern and colors of the broken tulip could only be replicated by taking a growth from the bulb, requiring an additional level of skill and knowledge. Through this manipulation of the infection, and cross-hybridization through proximity, seventeenth-century gardeners produced thousands of cultivars of tulips (Robinson 2009: 103–4).

By the eighteenth century, tulip cultivation had spread across Europe with specialists producing highly sought-after bulbs through skilful artifice. In the period from 1718–30, the Ottoman Empire embraced a more open stance in its relationship with Europe, referred to as *Lâle Devri* in Turkish, or the Tulip Era in English. Under the reign of Sultan Ahmed III (r. 1702–30), and his son-in-law, the grand vizier Nevşehirli Damat Ibrahim Pasha (r. 1718–30), the Tulip Era was a period of more peaceful relations with Europe, increased extravagance and courtly spectacles for urbanites, and a Turkish episode of tulipmania. Lavish garden festivals were held at night with lanterns illuminating planting beds filled with tulips (Wishnitzer 2014: 515). Affluent consumers competed with one another to

FIGURE 0.6 Judith Leyster, *Tulip*, *c.* 1643. Frans Hals Museum, the Netherlands. Image courtesy of Sailko, Creative Commons Attribution 3.0 Unporte.

distinguish themselves through horticultural display. Prices for bulbs rose dramatically in 1726 as consumption outstripped local supply, and government regulation of sellers was necessary to prevent profiteering and unrest (Salzmann 2000: 94). Once an exotic export, tulips and hyacinths were imported from Holland to satisfy the demand for flowers in Turkey (Demiriz 1993: 57–8).

Poets and travel writers alike drew inspiration from the sweet smells of the flowerbeds that lined the waterside villas of the Bosporus, which were filled with tulips, roses, hyacinths, carnations (*Dianthus caryophyllus*), narcissus, cyclamen, and jonquil, as well as common field flowers. Lady Mary Wortley Montagu wrote to Alexander Pope from Edirne in 1717 that

> the whole ground is laid out in gardens, and the banks of the rivers set with rows of fruit trees, under which all the most considerable Turks divert themselves every evening, not with walking, that is not one of their pleasures, but a set party of them choose out a green spot where the shade is very thick and there they spread a carpet on which they sit drinking their coffee ...
>
> (Montagu 1993: Letter XXXI).

These remarks contrast with the images of Europeans in gardens where walking was the primary diversion. A nobleman at the court of Louis XIV, Louis de Rouvroy, the duc de Saint-Simon (1675–1755), wrote of nature "subjugated by art" at the gardens at Versailles:

> The gardens astonish by their magnificence, but cause regret by their bad taste. You are introduced to the freshness of the shade only by a vast torrid zone, at the end of which there is nothing for you but to mount or descend; and with the hill, which is very short, terminate the gardens. The violence everywhere done to nature repels and wearies us despite ourselves
>
> (Saint-Simon 2006: ch. LXXII).

For the urbanites in Constantinople and Adrianople, flowerbeds were not a sequestered patch of nature to be appreciated in passing, but a space of repose and leisure, filled at dusk with lanterns for night-time celebrations.

The image of Turkish courtiers lounging in tulip beds, like Sybarites, was not embraced at all levels of Ottoman society. The luxury of the gardens occurred in an environment of increased elite consumerism and a perceived rise in European cultural influence, leading to false attribution of style and cultural exchange, especially in architecture and garden design. This perception of Western influence and conspicuous consumption led to popular riots, and the Anti-Tulip Rebellion in 1730. The rebellion claimed the life of Ibrahim Pasha and Ahmed III was deposed. The sumptuous pleasure gardens were destroyed, the palace at Sa'adabad was demolished, and courtiers were forced to burn their estates to the ground (Salzmann 2000: 97). The symbols of excess and luxury were destroyed during the rebellion, and the new regime pledged a pilgrimage as an act of contrition. In an era named for the cultivation and display of plants for pleasure, the Sultan and Grand Vizier neglected the cultivation of the state for the good of the people, and the Tulip Age came to an end.

PRIVATE COLLECTORS AND AMATEUR BOTANISTS

Outside of royal spaces dedicated to the symbolic expression of power through dominion over nature, individual collectors and cultivators attempted to manipulate, control, and classify plants. Many of these efforts were driven by the development of the philosophy of empiricism and the rising influence of the scientific method. There were some commercial aspects of these new developments, but others viewed botany as a pure science to be appreciated for its own value. The introduction of exotic plants to Europe was a challenge pursued by private collectors and amateur botanists who competed successfully with the established state-sponsored horticultural institutions and joint-stock companies in the collection of foreign plants and the development of innovative growing techniques.

Cultivating rare plants in Europe, especially citrus and evergreens, required specialist buildings called orangeries, where plants could be moved indoors during winter. While royal patrons had grand orangeries as part of their estates, more modest collectors of plants with botanical interests sought the advice of experts, who distributed their knowledge through new books on the subject and their extensive networks of contacts. Publication often supported specialist trade in rare seeds and saplings, that explorers and merchants would obtain and sell on not only to garden patrons, but also to apothecaries and hospitals. Jan Commelin (1629–92) was a Dutch merchant and botanist, who specialized in exotic plants, especially those imported from the East and West Indies and the Cape of Good Hope. In his *De Nederlandze Hesperides* (1676), Commelin provided advice on how to set up an orangery with an open court for display in summer months. Further technical innovations in the 1680s contributed to the development of tropical

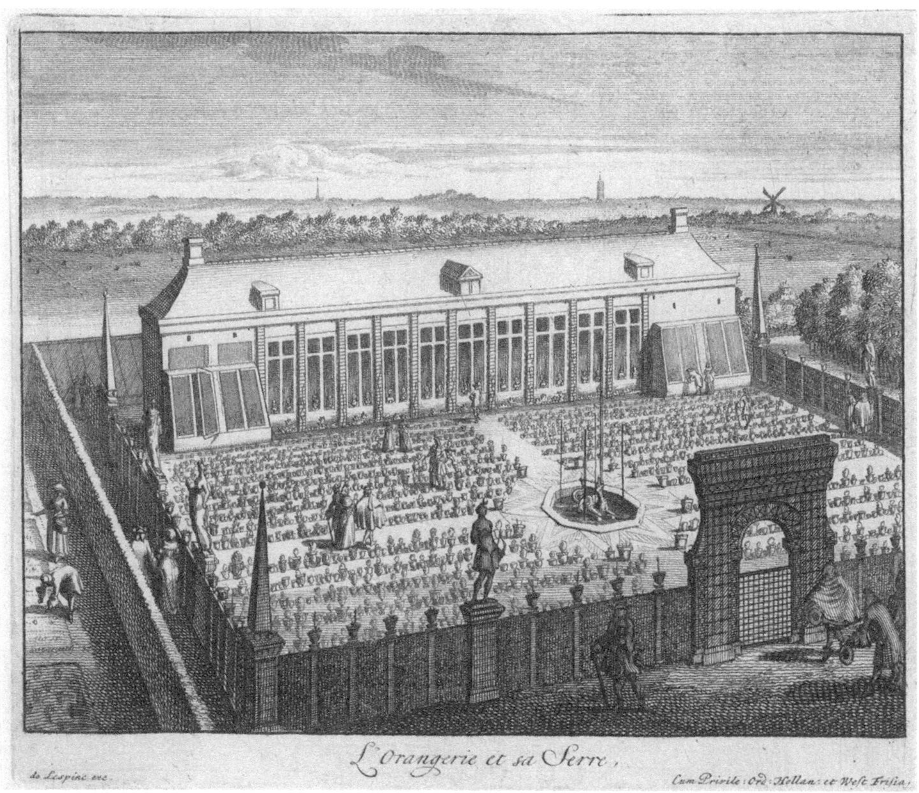

FIGURE 0.7 Joseph Mulder and Willem Swidde, *Orangerie van Kasteel Gunterstein, Breukelen L'Orangerie et sa serre* (1680–96). Rijksmuseum, Amsterdam (https://www.rijksmuseum.nl/en/collection/RP-P-1899-A-21594). Public domain.

hothouses that made it possible to grow more varieties of tropical plants. His second cousin, Magdalena Poulle (1632–99) developed networks with Dutch merchants and botanical collectors in order to establish her own collection of exotic plants, cultivated in an orangery with state-of-the-art hothouses in her extensive gardens at her manor house, Gunterstein, in the province of Utrecht.

Poulle was visited by plant collectors, botanists, and garden enthusiasts from other parts of Europe. John Evelyn (1620–1706), an English writer and gardener, listed Poulle's gardens as one of the most famous in the Dutch Republic and Henry Compton (1632–1713), the Bishop of London, sent his head gardener George London there to report on gardening innovations and the cultivation of exotics. The inventory of forty-four rare and remarkable plants in the gardens of Poulle listed by London included sugar cane (*Saccharum officinarum*), carob (*Ceratonia siliqua*), papaya (*Carica papaya*), tamarind (*Tamarindus indica*), hibiscus, delphinium, a coconut palm (*Cocos nucifera*), "*Fritillaria crassa*" (a variety then in cultivation in the Netherlands), *Sesbania grandi flora* from Southeast Asia, a coral tree (*Erythrina* sp.) from Brazil, a *Clematis* from Argentina or Paraguay, "*Leonurus Afric. Breyn*," thought to refer to a species of *Phlomis* from the Cape of Good Hope introduced into Europe in 1663, and "*Thlaspi sempervirnens et florens*," identified as *Iberis* from Persia (Sikkens-De Zwaan 2002: 216). All of these plants required special conditions for successful cultivation and the collection earned Poulle considerable fame. She enjoyed displaying her exotic plants in pots laid out on terraces in the summer months, complemented by fountains that were extravagant for private gardens of the merchant class. After Poulle's death in 1699, the plant collection was sold at auction by her brother, with the advertisement listing orange, lemon, myrtle, jasmine, camphor, arbutus, and oleander trees, and then referring to strange foreign tree fruits, roots, and bulbs collected over many years from distant parts of the world (Sikkens-De Zwaan 2002: 216–18).

PLANTING THE SEEDS OF ENLIGHTENMENT

By the time Poulle's plant collection was dispersed at auction, European societies had already begun to shift away from the authority held by the church and the monarchy in the seventeenth century to a more secular emphasis on empiricism and reason in the eighteenth century. The rise of science and philosophy, coupled with increased maritime trade and exploration led to new ways of understanding the place of humankind in the universe, and new ways of viewing and understanding nature. This change can be seen developing in early seventeenth-century English botanical texts such as John Parkinson's *Paradisus in Sole Paradisus Terrestris* (1629) and *Teatrum botanicum* (1640). A literal translation of the title of his first book is "Park-in-Sun's [a pun on the author's name] Terrestrial Paradise," which he followed with the "Botanical Theatre," referencing an earthly Eden and the Shakespearean metaphor of the world as a stage. *Paradisus* was an important botanical catalogue that categorized, described, and depicted over one thousand plants (Henrey 1975). The illustrated folio volume was a useful resource with over 780 original images on 109 woodcut plates by Christopher Switzer to complement the descriptive entries and information on gardening practices (Anderson 1997).

In his "Epistle to the Reader," Parkinson extolled the universal virtues of plants and their study: "he is not human that is not allured with this object. The study, knowledge,

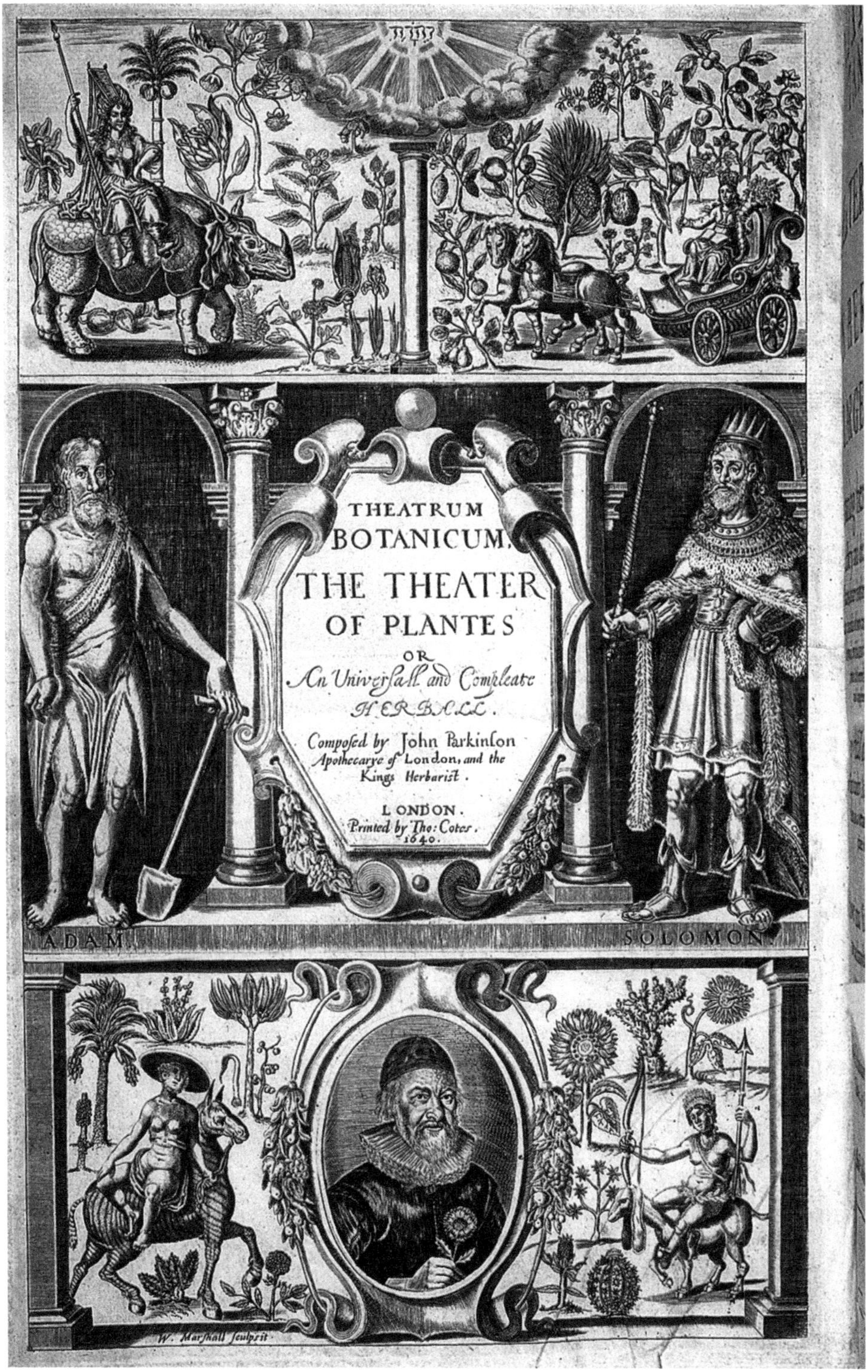

FIGURE 0.8 John Parkinson, *Theatrum Botanicum. The Theater of Plantes*, 1640. Wellcome Collection. Attribution 4.0 International (CC BY 4.0).

and travel in them ... is not only pleasant, but profitable, by comforting the mind, spirits and senses with a harmless delight and by enabling the judgment to confer and apply help to many dangerous diseases." Plants are not only decorative or appealing to the spirit and senses, but also have utility through their medicinal value. Parkinson was an expert gardener himself, with two acres and an estimated collection of nearly five hundred plants at Long Acre, in West London (Riddell 1986). He obtained new specimens through an international network of explorers, botanists, and horticulturists. Long Acre contained hundreds of plants from overseas, mainly western Europe, but some were sourced from North Africa, North America, and the West Indies. The networks that facilitated the international movement of plants and seeds had a parallel in the intellectual networks that promoted the circulation of ideas that continued to grow during the Enlightenment.

In addition to introducing exotic plant species to England, Parkinson was a founding member of the apothecary society and royal apothecary to James I (r. 1603–25). *Theatrum botanicum* provided details on the healing properties of plants found in English medicinal gardens, such as the one established at the Chelsea Physick Garden in 1673 by the Worshipful Society of Apothecaries. Parkinson's systematic organization of the text into seventeen "sundry classes, or tribes" was a rudimentary attempt at classification before taxonomy had been developed. He drew on decades of professional expertise to offer empirical judgment based on "many years travail, industry, and experience" in contrast to the "many errors, differences, and oversights" that he argued had been offered by less seasoned authors. The emphasis on observation and analysis sets the text apart from earlier treatises based on repetition and conjecture.

Through his combination of discourse and practice, Parkinson enlarged Early Modern knowledge of plants. *Paradisus* communicated his advances in the introduction and acclimatization of exotics and *Theatrum Botanicum* became the standard herbal for generations of English apothecaries. Together, these texts addressed both the practical world of domestic pharmacology and the cultivation of domestic and exotic plants with an encyclopaedic and scientific approach based on observation and experience.

EMPIRICISM

The primacy of observation and experience was central to the thinking of the group of Early Modern philosophers, the English empiricists, who held that all human knowledge of the external world originates from sense experience. This philosophy was articulated in John Locke's (1632–1704) influential treatise *An Essay Concerning Human Understanding* (1690). Locke's formulation of empiricism influenced botanists through his discussion of the "natural" system and the key question of classification. His essay used more than seventy references to plants as examples to support and explain his reasoning (Cain 1997). True to his philosophy, Locke drew on his substantial interest in botany and his extensive collection of plant specimens to inform his views on the nature of species and the theory of classification (Anstey and Harris 2006).

Locke employed the transformation of an oak (*Quercus robur*) from sapling to tree in a discussion of essence and identity, which was added to his second edition of the *Essay* in 1694. The chapter addresses the complex principle of personal identity beginning with more basic examples to establish definitions and principles for the reader. Locke used the example of the oak to consider whether an organic object is the same mass of matter or the same organism over time: "An Oak growing from a Plant to a great Tree, and then lopp'd, is still the same Oak" (ch. XXVII, section 3). He continued his analysis, addressing the increasingly complex identities of plants, animals, and human beings in turn. For the oak, he considered the difference between the "cohesion of particles" and the "organization of those parts as is fit to receive and distribute nourishment so as to continue and frame the wood, bark, and leaves ... in which contains the vegetable life." That Locke chose the "vegetable life" of the oak to represent the persistence of identity while undergoing substantial organic change reflects the cultural familiarity of plants for the contemporary reader and the author's awareness of the same.

Locke chose a more exotic plant, the pineapple (*Ananas comosus*), to illustrate his argument that sense experience is the only real source of knowledge about the external world (Silver 2008). Pineapples in the late seventeenth century were rare and expensive in England, so many readers would not have had the means or opportunity to taste one. Instead, any idea of the taste would be second-hand from an acquaintance, or through a written description from a traveler or West Indies merchant. Locke chose a deliberately unusual plant that readers would know, but most likely not have any first-hand experience. Unlike the familiar metaphor of the oak, the novelty and rarity of the pineapple was employed to show how descriptions, and ideas based on discourse alone, were a poor substitute for sensory experience of it.

CULTIVATION OF THE INDIVIDUAL

Locke was not the only philosopher who was also skilled as a botanist. Jean-Jacques Rousseau (1712–78) was an influential Swiss philosopher who, like Locke, developed a taxonomy for plants, was an active collector of plants, and part of an international network of collectors and scientists who exchanged specimens. He also contemplated a system of classification for plants which placed more emphasis on a range of morphological features rather than an exclusive focus on the reproductive system (Cook 2000). Rousseau's writings on philosophy and education were informed by his pursuit of botany and his belief in the primacy of nature (Cook 2012).

The manipulation of nature by man was depicted by Rousseau in his educational novel *Emile* (1762) as a perversion that degraded plants and humans alike: "Everything is good leaving the hands of the author of all things; everything degenerates in the hands of man. He forces one soil to nourish the products of another ... he disfigures everything; he loves deformity, monsters. He wants nothing as nature made it, not even man ... man must be fashioned in keeping with his fancy like a tree in his garden" (Rousseau 1762: 162). For Rousseau, the study of botany was an end in itself, not a discipline dedicated to practical applications for the gardener or apothecary (Kleinau 2012: 474). His love of plants in their natural environment, however, did not translate into an educational philosophy that suggested man should be left untended, in a natural state. The uncultivated person would lose their nature through exposure to the pressures of society: "Plants are fashioned by cultivation, men by education" (Rousseau 1762: 162). Nonetheless, Rousseau pointed out that there were limited benefits to using manipulation to direct growth in a specific direction, like an espaliered tree: "Such, for example, is the habit of the plants whose vertical direction is interfered with. The plant, set free, keeps the inclination it was forced to take. But the sap has not as a result changed its original direction. And if the plant continues to grow, its growth resumes the vertical direction" (Rousseau 1762: 163). Men, set free like the plant, resume their inclinations when new growth occurs. Rousseau, himself, enacted this principle in his philosophical musings on plants in his later years. *The Reveries of a Solitary Walker* (1776–8), Rousseau's last work, is a justification of his way of life, including his radical political and religious views that led to exile and isolation (Cantor 1985: 363). Deeply autobiographical, Rousseau describes walks around Paris, gazing at plants. It is through his engagement with botany that Rousseau reorients his philosophy of careful examination of the world—both that of nature and of man.

Like Rousseau, the German writer Johann Wolfgang von Goethe (1749–1832) believed in the therapeutic benefits of botany and its role in personal development, or *Bildung* (self-cultivation). He proposed meditating on the inner workings of plants, which would lead to an awakening of the senses. In his pursuit of science, Goethe formulated an alternative, more natural, system of plant classification and rejected the Swedish botanist, zoologist, and physician Carl Linnaeus' (1707–78) artificial taxonomy (Larson 1967: 592). Instead, he proposed a research method based on a combination of information received through the senses with imagination and intuition. This approach would allow researchers a glimpse into the inner nature of plants rather than the study of the individual parts of plants. His philosophy was particularly focused on transformation and interdependence in plants, something he described as "the truth about the how of the organism," and emphasized holistic analysis (Goethe 1790). In particular, he encouraged the study of plant morphology as a method to better understand the natural world.

PLANTS IN A PHILOSOPHER'S GARDEN

The desire to have a better understanding of the natural world included the study of plants in discourse and in practice. For some writers there was little separation between the metaphorical and tangible cultivation required for the improvement of civilization. The overlap between social criticism, political reform, and practical horticulture arose repeatedly in the works of botanists, novelists, and political thinkers. In 1659 in a letter to Anglo-Irish chemist and physicist Robert Boyle (1627–91), John Evelyn urged "everyone to cultivate his own garden" (Ros 1981: 235). He proposed the creation of a model society which would serve as an example of the better way of life in the corrupt world. This new paradigm would consist of nine persons working together to cultivate a "physick-garden, kitchen-garden, and a plantation of orchard-fruit" on a thirty-acre estate (Evelyn 1827: 84). Evelyn's letter was published in 1750, informing the work of French writer and thinker François-Marie Arouet (1694–1778), better known by his pen name, Voltaire. In his picaresque novel *Candide* (1759), Voltaire's main character finishes his epic journey with the purchase of a modest farm where he and a handful of comrades will work the land. Through their cultivation, the group avoids what Voltaire named as the three great vices: boredom, vice, and poverty. Candide's phrase at the end of the novel, "We must cultivate our garden," is both a retreat from the cruelty of the world and a call to arms. The necessity of the harvest drives the individual to work the communal land, making a tangible contribution for the benefit of the group. Productive work at a local level was a positive action taken to ward off the evils of the world and required each person to take responsibility for their actions.

Both Evelyn and Voltaire believed that horticulture was a key element of rural life, and for the philosopher, plants were regularly incorporated metaphorically into his political writings (Ross 1981: 235). But plants were not simply of metaphoric use to Voltaire. He was an avid gardener and cultivated the gardens at Les Délices, the house he rented on Lake Geneva (Lac Leman), while he became an activist for political reform, defending the rights of persecuted people (Hirvonen 2014: 232–3). Voltaire began a program to improve the gardens, immediately after taking up residence in 1755. To Jean-Robert Tronchin (1710–93), the owner of Voltaire's residence, he described his improvements:

> I have already planted 250 trees; I have created some avenues for you ... your brother has just brought me some seeds; it was the nicest present you could give me. At this moment I am sowing your Egyptian onions; even the Israelites did not like them more than me. Please send me everything you can in the way of flowers and vegetables. The garden was completely bare; we must start from scratch I am founding a second Carthage.
>
> (quoted in Davidson 2004: ch. 3, n23)

Voltaire's reference to Carthage recalls the formidable agricultural production of the city, and his reference to Egyptian onions links his garden to the ancient plants of the Bible. A week later, he ordered a range of plants and seeds to produce fruit, vegetables, and herbs for the table: globe artichoke (*Cynara cardunculus*), lavender (*Lavandula angustifolia*), thyme (*Thymus vulgaris*), rosemary (*Salvia rosmarinus*), mint (*Mentha* spp.), basil (*Ocimum basilicum*), rue (*Ruta graveolens*), strawberry plants, tarragon (*Artemisia dracunculus*), sage (*Salvia officinalis*), and hyssop (*Hyssopus officinalis*) "to cleanse our sins" (Davidson 2004: ch. 3, fn24). A medicinal herb of ancient origin, hyssop is mentioned in the Bible several times and was associated both with the ceremonial cleansing of people and houses. In the seventeenth and eighteenth centuries, the leaves of hyssop were used in a tea to treat dropsy (edema). Even Voltaire's shopping lists for plants are filled with allusions to Classical antiquity, the Bible, and the staples of *materia medica*, blending philosophy, literature, history, and the practicality of herbal medicine in the most simple of texts. The enthusiasm with which Voltaire pursued the establishment of his garden at Les Délices demonstrates his firm belief in the benefits of plant cultivation, something he shared with Rousseau and Goethe. That this should coincide, in Voltaire's later years, with his most fervent political activism, is more than coincidental, it is evidence of his personal philosophy in action on a both a literal and metaphorical level.

PHYSIOCRACY

The practical implications of Voltaire's call for all to cultivate their own garden took root in the late-eighteenth-century political economy of French authors like François Quesnay (1694–1774) and the Marquis de Mirabeau (1749–91), in a movement called physiocracy. The physiocrats held that the true source of national wealth was measured only in surpluses produced by agriculture. In France, the central role of agricultural productivity was undermined by internal tariffs, poor infrastructure, monopolistic guild regulations, and restrictions on labor migration. All of these governmental interventions were claimed to be against the "natural order" proposed by the physiocrats who called for deregulation and a single tax, an economic policy of laissez-faire (Hochstrasser 2006: 419). While the physiocrats were based in France, their influence of their political economy extended to Italy, Spain, Bengal (India), and the newly established United States. Physiocracy was promoted by Jefferson in the United States and can be seen in the Republican Party's support for the ideology of agrarian democracy (Albertone 2009: 123).

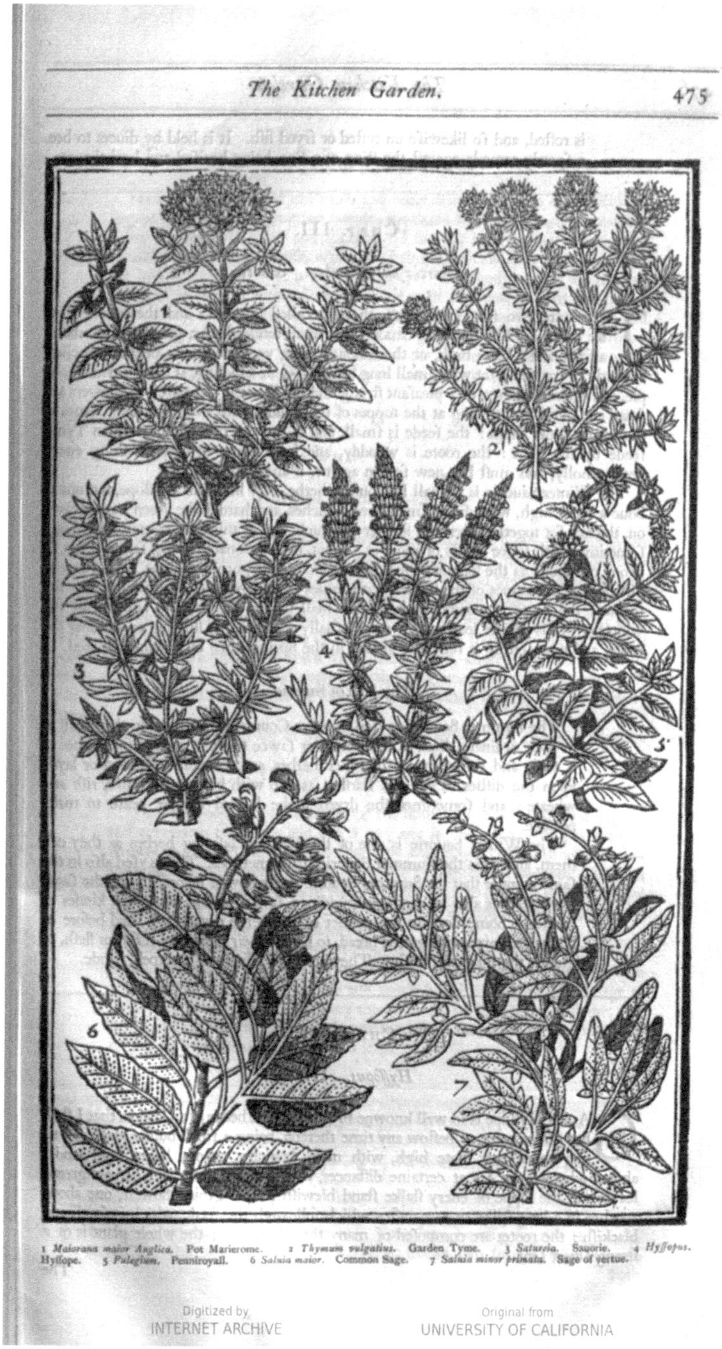

FIGURE 0.9 An engraving depicting Garden Herbs: Thyme, Marjoram, Savoury, Hyssop, Pennyroyal, and Sages. From John Parkinson, *Paradisi in Sole Paradisus Terrestris* (London, 1629). Hathi Trust. Public domain.

Jefferson's "Summary of Public Service," written sometime after September 2, 1800, questions what an individual contributes to a nation's history: "I have sometimes asked myself whether my country is the better for my having lived at all? I do not know that it is." But to answer this question, Jefferson pondered what he has "been the instrument of doing" and lists weighty political documents such as the Declaration of Independence and acts concerning citizenship and natural rights, as well as thoughts concerning prisons and public schools. Plants figure as well. He noted the important of olive plants (*Olea europaea*) from Europe which, as a result, flourish in the southern states as "the germ of that culture." He also recalls "a cask of heavy upland rice" from Africa, sent to the South "in the hopes it might supersede the culture of the wet rich which renders S. Carola & Georgia so pestilential through summer." He concluded "the greatest service which can be rendered any country is to add an useful plant to its culture; especially a bread grain. Next in value to bread is oil" (Jefferson 1800).

Experimentation with plants was a preoccupation of Jefferson throughout his adult life, as confirmed through his notes kept in his garden journal from the 1760s until his death in 1824. In this diary he recorded, measured, and observed the natural world. Published with Jefferson's other writings on plants, including many of his letters, the Garden Book is one of the richest sources of Enlightenment experimental plantings in the New World (Betts 1944). Jefferson's plantation estate at Monticello outside of Charlottesville, Virginia, included some three hundred cultivars of ninety-nine species of herbs and vegetables (Hatch 2012: 4). In addition to the vegetable garden, Jefferson had orchards and surrounding vineyards, as well as a flower garden close to the house. These plantings, for utility and pleasure, were interconnected to Jefferson's idea of himself as a man of the Enlightenment.

FIGURE 0.10 Jane Braddick Peticolas, *View of the West Front of Monticello and Garden*, 1825. Watercolor on paper, 34.6 × 46 cm. Monticello, Charlottesville, Virginia. Public domain. Wikimedia Commons.

LIFE AT MONTICELLO

Jefferson's engagement with plants for diverse purposes over many decades presents as a microcosm of the cultural experience of plants at the end of the eighteenth century. Plants were fundamental to Jefferson's understanding of, and interest in, the natural world, and the example of his life at Monticello sets up the themes that are explored in depth in the individual chapters of this volume. He developed an experimental garden to deliver staple foods (Chapter 1), not only to his own family, but with an eye towards sustaining the people and economy of a new nation. His efforts in agriculture including a desire to promote the cultivation of olives in the southern states, writing to William Drayton (1776–1846) of Charleston in 1789, "I have exceedingly at heart the introduction of [the olive] into Carolina & Georgia being convinced it is one of the most precious productions of nature and contributes the most to the happiness of mankind" (Cornett 1995). The enslaved chef and other workers in the kitchens of Monticello were trained in French cuisine, and transformed the plants grown in the gardens into luxury foods (Chapter 2) for Jefferson's table, as an expression of enlightened hospitality (Craughwell 2012).

Jefferson used his networks to trade in plants (Chapter 3), both exporting plants to Europe and importing plants through suppliers and friends, including the self-taught botanist and naturalist John Bartram (1699–1777) and his sons. Founded in 1728, Bartram's Garden in Philadelphia was one of the most influential resources of its kind in mid-century America (Cornett 2002; Hatch 2007; Myers 2011) and remains the oldest extant botanical garden in the United States. Botanical discovery was one of the explicit purposes of the Lewis and Clark Expedition, initiated by Jefferson and sponsored by Congress in 1803, with a first shipment of over a hundred specimens, including the "Mad Dog Plant" (*Echinacea angustifolia*) sent back from the explorers' winter quarters at Fort Mandan Missouri in 1805 (Hatch 2003). Jefferson was fascinated not only by the plants themselves, but with their practical application. Many years earlier, when he was drafting his *Notes on the State of Virginia* (1785), Jefferson's only published book, he organized native plants into four categories according to their function: "Medicinal," "Esculent" [edible], "Ornamental," or "Useful for fabrication" (Chapter 4). At the same time, he regularly used plants metaphorically for the promising future of the New World (Chapter 6).

As he was skeptical of medicine as practiced at the time, Jefferson promoted the natural healing power found in nature (Chapter 5). His neighbor, Lucy Meriwether Lewis Marks (1752–1837, mother of Meriwether Lewis of the Lewis and Clark Expedition) was known locally as a "yarb," or herb doctor, and had extensive knowledge of the medicinal value of herbs as healing plants. Jefferson used Peruvian bark (*Cinchona calisaya*) to relieve his headaches and jalap, a purgative drug obtained from the tuberous roots of a Mexican climbing plant [*Ipomoea purga*], to treat his gout.

Plants were part of the ornamentation of Jefferson gardens (Chapter 7), especially the flower garden that extended from his house. He grew approximately 105 kinds of herbaceous flowers, although he admitted to a friend that he could not recall if he had ever planted a flower, leaving this work to his daughter and granddaughters (Hatch 2005). Caring for the flower gardens was one of the primary duties for women in the Jefferson household, considered a suitable occupation for ladies. As amateur artists, they also learned to draw plants and represent them (Chapter 8). In sum, as an aspect of daily life at Monticello, plants figure into every aspect of Jefferson's cultural understanding of his place in the world.

CONCLUSION

Exploration and colonization fundamentally altered the role of plants in Enlightenment culture as they became something to be studied and admired, collected and categorized, traded and exchanged, experimented with, and consumed. Moving from Vaux-le-Vicomte to Monticello, it is possible to see the myriad ways that plants were valued and understood during the seventeenth and eighteenth centuries: for functional purposes, as items of discovery and exchange, as products to be transformed, as useful metaphors, for aesthetic embellishments, and as objects of desire. This sets up the more detailed analyses that attend to these specific aspects of the cultural history of plants during the period in the chapters that follow.

CHAPTER ONE

Plants as Staple Foods

JANE LEVI

Robert Herrick's poem "The Hock Cart, or Harvest Home," part of his *Hesperides* collection published in 1648, provides the food historian with plenty of lyrical material with which to embellish descriptions of the bucolic pleasures of harvest home in seventeenth-century England (along with a number of hints as to the menu for the traditional supper served on such occasions):

Come Sons of Summer, by whose toile,
We are the Lords of Wine and Oile:
By whose tough labours, and rough hands,
We rip up first, then reap our lands.
[...]
And, you must know, your Lord's word's true,
Feed him ye must, whose food fils you.
And that this pleasure is like raine,
Not sent ye for to drowne your paine,
But for to make it spring againe.

(Herrick 1648: 115)

Its underlying message to Mildmay Fane, 2nd Earl of Westmorland (1602–66), to whom it is dedicated, also operates as something of a political touchstone for the period discussed in this volume. Herrick's exhortation to the landowning classes to appreciate the toil of their field laborers, since it is by their work that they themselves are both fed and enabled to purchase their own edible luxuries of wine and oil, echo social concerns that both predate and continue to foment from the mid-seventeenth into the early nineteenth century. During this period, in Europe in particular, ideas about land ownership and management, agricultural techniques and "improvement" underwent significant, often revolutionary change. At the same time, the continuing development of cities, the expansion of empires, and accelerated population increase brought to the fore concerns about how to feed large numbers of people living at subsistence levels but without the direct means to subsist through their own cultivation of land. The responses to these challenges, and the changes wrought in the social and economic landscape, including patterns of international trade within empirical structures, had a significant impact on the agricultural landscape at home. A consideration of staple foods, agriculture, and thus the land, lies both literally and figuratively at the root of the matter.

FIGURE 1.1 *A celebration party given in honour of a good harvest*. Engraving by B. Picart, 1733, after Virgil. Wellcome Collection. Attribution 4.0 International (CC BY 4.0).

"ONLY A STOMACH THAT SELDOM FEELS HUNGER SCORNS THINGS COMMON" (HORACE, *SATIRE* 2.2.37)

In this examination of plants as staple foods the chapter is most concerned with the plants that formed the daily meals of the majority of people, and those that contributed to the production of foods seen as essential to maintenance of a healthy, sufficient, and culturally acceptable diet. This becomes a really interesting question in this period, one of so much change. In various parts of the United Kingdom, as well as the rest of the world, there were major shifts in the sources of particular staple plant foods, as well as the introduction, or at least extensive promotion, of new ones. At the same time, many of the core elements of diet remained notably stable, both within cultures and across social divides. While the Lords in Herrick's poem have their luxuries of imported wine and oil, they share their workers' dependence on many of the underlying staple grains and vegetables that they are growing; those basic foodstuffs fill their bellies, too, even if the dishes they make from them and the quality of the ingredients they have access to might vary. As is known from food scholarship, culture determines much of what a person accepts as being edible, or constituting real food. It is remarkably difficult to change a person's ideas about what, for them, counts as "food" or "a meal," wherever they fit within a social hierarchy. So, the challenge is never as simple as just feeding people; the question is how to feed them the food that they accept as such. Nowhere is innovation less immediately acceptable than in the basic staples of life, yet this period was one of immense upheaval and developments that challenged deep-seated beliefs about the social order, including its relationship with food.

The first part of this chapter provides some agricultural context, followed by a general introduction to the key principles of and staple foods in the seventeenth- and eighteenth-century diet. It goes on to discuss the main component plants consumed in the UK and parts of Europe, touching on production and consumption of the foods by different strata of society. The focus is on grains, pulses, and vegetables. Very brief comments are given on the rest of the world, with some suggestions for further reading.

IMPROVEMENT

As the middle classes burgeoned, this period in Europe saw the beginning of an agricultural revolution, seen in retrospect as the necessary precursor to the industrial revolution. It began with the idea of "improvement," a theme pursued by numerous writers in Britain. "Improvers" called for transformations in agricultural techniques and land use, focused on making existing farmland more productive and profitable, and finding methods of making previously uncultivated land fruitful, preferably within a very short space of time. As Joan Thirsk (1922–2013) reports it, "in common parlance in the first half of the seventeenth century, improvement meant putting land under the plough: arable was still conventionally deemed a more advantageous use of land" (1990: 139). A pamphlet of 1653 describes the wild and waste (i.e. uncultivated) land across the country as "a deformed chaos" that was "a shame and a reproach" (Thirsk and Cooper 1972: 135) to the people of England. The "wild howling wildernesses" should be enclosed, tilled, manured, and planted according to the soil type, so that they:

> would bring forth plenty of flax, hemp, hops, corn, also increase cattle of all kinds, and many other things by which means the State would be sufficiently supplied with hemp

for cordage for their shipping, and the poor more richly replenished with bread corn (the staff of sustenance) and many other necessary and profitable fruits.

(Thirsk and Cooper 1972: 135–6)

Improvement would supply plenty, enriching the state and feeding the poor with bread, their staff of life. This language of the improvers, who consistently point out the untapped potential of England's uncultivated land, was borrowed by their contemporaries but political opposites, the Diggers, to justify their occupation and cultivation of common land in 1649. The Diggers' approach was radical within an existing political structure that categorically ruled out the use of common land for cultivation: "Divide *England* into three parts, scarce one part is manured. So that here is land enough to maintain all her children, and many die for want, or live under a heavy burden of povertie all their daies" (Winstanley 1649: 61). The Diggers' declared intent was to exercise their capabilities to improve the common land for themselves, meaning the common people as a whole, while the improvers generally assumed improvement was contingent upon mass enclosure and thus the leadership and control of landowners. In the introduction to his *English Improver*, Walter Blith (1605–54) points out that an improver's success is not a foregone conclusion, and that hard work is required: "Study Industry, Improvement is neither Father nor Mother unto plenty, but I may say it is the Midwife that Facilitates the birth" (1649: a3). Acknowledging the influence of earlier writers on husbandry such as Gervase Markham (*c.* 1568–1637), Thomas Tusser (*c.* 1524–80), and Bacon's *Natural History*, Blith's treatise went on to explain how his proposed experiments should be conducted, insisting that demonstration—which he himself had not yet carried out—would prove the inevitable success of his techniques (Thick 2014: 12).

One of Blith's acknowledged inspirations in his later work was Samuel Hartlib (*c.* 1600–62), who had set up and run the "Office of Address" in 1646 following the Parliamentary victory in the English Civil War (Thick 2014: 19). This office pulled together and disseminated useful ideas from the so-called "Hartlib Circle," with a strong focus on new ideas about husbandry drawn from all over Europe, particularly the Netherlands (Thick 2014: 19–20). Hartlib's short 1641 fiction, *A Description of the famous Kingdom of Macaria*, focused on practical ideas for a kind of "light-touch" government, required to operate only occasionally in order to administer a system of five under-councils, in turn structured to exploit and share the bounty of agriculture (Husbandry), Fishing, Trade by Sea, Trade by Land, and New Plantations. Indeed, Hartlib's ideas about husbandry as one of the primary governing forces in the fictional Macaria reflect the immense energy he devoted to the subject in the real world, and emerge again with factual force a decade later in his proposal to Parliament to found a college of husbandry (Hartlib and Dymock 1651). Although Hartlib's work consistently referred to the benefits to the poor of making the land more productive, in his 1648 work, *A further Discoverie of the Office of Publick Addresse for Accommodations*, he suggests the benefit will mainly accrue to them by providing supervised work under an improving master (Hartlib 1648: 7). Increasingly in the 1650s, he and his major correspondent Cressy Dymock (*c.* 1619–70), like Blith, directed their focus to the work to be done by enclosers and existing landowners (Hartlib and Dymock 1653). While in part this was said to be for the benefit of the so-called "deserving" poor, there was no real suggestion of self-directed activity by the poor

FIGURE 1.2 *Farm workers under a tree are making a basket, a plough and other agricultural implements*. Engraving by Wenceslaus Hollar after F. Cleyn, 1654. Wellcome Collection. Attribution 4.0 International (CC BY 4.0).

themselves. The language of improvers focused almost solely on the economic and social benefits of bringing land into productive use within the existing social structure, made no assumptions about downward changes in its distribution, and emphasized hard work and skill as the prime factors contributing to success. These early contributions to the literature set the tone for the rest of the period under discussion here, in which the pattern of top-down control of production continued and, in the case of commons use and enclosures of common land, accelerated.

Besides providing opinion on the need to transform wide swathes of the country, books on improvement also focused on good husbandry. This advice was aimed more directly at a middle-class readership, providing for themselves a proportion of the "garden stuff" they expected in their daily diet. It is clear from the short book or seed catalog published by the London seed merchant Steven Switzer (1682–1745) in 1728 that seeds from Italy for cardoons (*Cynara cardunculus*), new brassicas, and other fine vegetables could be ordered for delivery along with fine silks and other luxury goods arriving from the continent (Switzer 1728), a long-established pattern. Knowledge of husbandry included knowledge about the right month and appropriate phase of the moon for the sowing of seeds, planting of seedlings, and gathering of seed for the next planting (Markham 1615a). The lists and tables that appear in manuals in the Early Modern period—including Culpeper's catalog of herbs and plants assigned to their planets—bear witness to a cosmological imagination that has ancient roots as well as its own hermetic tradition (Culpeper 1652). Indeed, the biodynamic calendars followed by some contemporary specialist growers (who also publish evidence of the difference in performance of seeds sown in the "wrong" moon phase versus the "right" one within the same year and season [Thun and Throll-Keller 1999]) bear a striking similarity to the Early Modern deployments of these kinds of information. It seems clear that these details were well known at the time, so that works on husbandry like Gervase Markham's *Countrey Contentments* (including *The English Huswife*) and his *Farewel to Husbandry* (in *A Way to Get Wealth*), which went through repeated printings from 1615 until well into the following century, were recording and codifying common knowledge as much as they were bringing new knowledge to those who did not already have it. More encyclopedic botanical works like John Parkinson's *Theatrum Botanicum* of 1640 and Robert Lovell's *Enchiridion Botanicum* of 1659 gave the planting months for all of the plants listed. Assuming one could get (or, more accurately, afford) seed, it is therefore apparent that seed collection, sowing and, to a greater or lesser extent, harvesting possibilities existed in every season. Where seed was not available, this planting information gives an indication of what wild food (and seed) sources may begin to be available as the seasons progressed. Markham's list, for example, suggested a wide range of vegetables for spring planting:

> In March the Moone new sow Garlicke, Borrage, Buglosse, Chervile, Coriander, Gourds, Cresses, Marjoram, White Poppy, Purslan, Radish, Sorrell, Double Marigolds, Time, Violets. The full moone; Anisseeds, Bleets, Skirrits, Succory, Fennell, Apples of Love and Mervaylous Apples. At the wane, Artichokes, Bassill, Blessed Thystell, Cole Cabadge, White Cole, Greene Cole, Citrans, Cucumbers, Harts Horne, Samphire, Spinage, Gilliflowers, Issop, Cabadge, Lettice, Mellons, Mugrets, Onions, Flower Gentil, Burnet, Leekes, and Savory. In May the moone old sow Blessed Thistle.
>
> (Markham 1615b: 37)

DIET

It is safe to assume that somewhere in between the two extremes of pre-modern Galenic thinking on medicine and food and radical new ideas about modern life and husbandry was a dietary middle ground where most people resided. Here, the fundamental dietary provision was cereals—wheat, rye or barley, or oats in more northerly regions, made into various products such as bread and beer—with beans, peas, and other vegetables (including roots) at the heart of the basic accompaniments. According to means, this was enhanced with dairy produce (milk being the cheapest and most basic, followed by butter, then cheese); or meat in small quantities where affordable (Thirsk 2007: 92). Most working people were probably "Carnivorous by choice, vegetarian of necessity" (Hessayon 2008: 16), and this hierarchical positioning of basic vegetables, bread, and beer at the bottom—as the staples—embellished with smaller amounts of dairy products and then meat, seems to form a consistent pattern that generally seems to reverse in proportion to wealth. The difficulties of changing people's diet and their fundamental ideas about food have been alluded to above. Health manuals of the early part of the period illustrate this for us. Thomas Moffett's *Health's Improvement*, originally written at the end of the sixteenth century, was published in 1655 and republished in 1746. By that time, there had been significant developments in medical practice, but the old thinking on the nature of foodstuffs and their effects on different classes of body retained currency. In listing "anti-acid" vegetables, including cabbage, garlic, onion, cress, leek, radish, nettle, and many herbs, as well as plants that might be preserved by candying like angelica (*Angelica archangelica*) and eryngoes (*Eryngium maritimum* and *E. campestre*), Moffett emphasized these differences, telling his readers that:

> Poor People, whose Food is principally of the farinaceous Kinds of Vegetables, and who eat but little Flesh Meat, are subject to these Disorders [pustules and fits], but would be much more so without the strong Exercise they generally use; for Exercise, as was before observed, by strengthening the animal Fibres, and promoting the Digestion of the Aliment, and Assimilation of the Chyle, prevents an acid Acrimony from being formed in the Juices.
>
> (Moffett et al. 1746: 16)

While medical practices developed, and ideas about diet evolved, there is little evidence that the fundamental model of consumption underwent radical change, other than when economic means allowed for innovation in ingredients, or dearth forced experimentation with different foodstuffs. Even in the latter case, the general drive was to find a substitute means of making the same foods, rather than changing the basic structure of the diet.

GRAINS

In April 1649, right at the beginning of our period, reports appeared of strange happenings on a section of common heathland in Surrey, England (Firth 1894: 211). A small group of men were seen burning parts of the heath, and then observed returning nine days later to dig and plow this section of the common land, ready for planting. This group and others like them became known as the Diggers, and although their radical occupation and cultivation of common land was short-lived,

FIGURE 1.3 *Kitchen Scene*. Chalk and wash drawing by Cornelis Dusart, 1687. From the Collection of Rita and Frits Markus, Bequest of Rita Markus, 2005. The Metropolitan Museum of Art, New York. Public domain.

after a long period of obscurity they have become a touchstone for many subsequent causes related to land access and self-sufficiency thanks to George Gooch's labeling of them as "English communists" in 1898 and their subsequent celebration as such by influential twentieth-century historians such as George Sabine (1880–1961) and Christopher Hill (1912–2003). For this chapter, they provide a perfectly timed starting point for a discussion of staple food plants. However history may have interpreted and re-interpreted their political motivations, their digging had at the heart of it a clear, fundamental purpose: "to grow corn for the succor of man" (Winstanley et al. 1649: 4). Grains (the "corn" referred to here) lay at the very heart of the basic diet. Until the late nineteenth century "corn" referred not to maize but to the main cereals in production: wheat, barley, rye, and oats. Maize, usually now referred to as corn, arrived in Britain in the late sixteenth century but was not in major production there until the early twentieth century. The climate is not ideally suited to its growth and maize grown in northern Europe is generally used for forage (animal fodder) rather than human consumption.

In seventeenth- and eighteenth-century Europe bread was the primary staple food, and the majority of the grain produced was used either in bread production or in various porridges and gruels. This is made clear in countless health manuals and cookery books, which not only gave directions for making bread and hard crackers or biscuits from grain flours, but also grain-based dishes. Flour itself or already-made bread and biscuit could be used to make puddings, to thicken or make more substantial soups and stews, or cooked in water and mixed with another liquid such as milk, beer, broth, wine, or water to make porridges and gruels, or other nourishing food for invalids or the elderly (Moffett et al. 1746: 27). The key role of flour and bread throughout this period is emphasized by the allowances given to army men, as described by John Buchanan who traveled to Ostend as doctor to a unit of soldiers in 1746. The soldiers were given an allowance of "one pound bread, half a pound salt butter, and one quarter pound cheese for dayly dyet, with two quarts small beer [4 pints], or half a pint brandy, for drink," while the officers had the same food allowance, with unlimited ship's beer (Kopperman 2012: 50). It is interesting to note that on board ship there was also an allowance of a peck [a quarter of a bushel, a volume measure whose equivalent weight varied by grain] of grain for the soldiers' horses, which were fed wetted hay and "corn" three times a day. Flour from the supplies was used as a paste to stop bleeding in the mouth or neck of the horses (Kopperman 2012: 51). Basic foods were also used as common remedies for humans. A poultice of white bread and milk or turnip was said to be soothing to injured shins, while gripes and purging were treated with white beer and toasted crusts with a scraping of nutmeg (Kopperman 2012: 55–6).

The staple grains of Europe, used not only in bread but many other porridge, pudding, and soup-like dishes, also remained wheat, barley, oats and rye, all of them commodities forming crucial components of trade. The consensus seems to be that in the UK between 1640 and 1750 the proportion of wheat grown increased significantly, barley output grew, and rye and oats started to decline (Chartres 1990: 194). In the first fifty years of the eighteenth century, as a result of agricultural improvements, the grain surplus grew by 20 percent, and it is assumed there was a similar impact on the production of pulses (Chartres 1990: 194). Cereals, the vast proportion of which were grown in East Anglia, were an important export commodity, with wheat being exported to Spain and Portugal, and to a lesser extent Italy, with France providing another important trading partner when the countries were not at war. Rye and malt oats were exported to Ireland, while significant quantities of barley were sent to Spain and to the Low Countries. A proportion of the grain (especially malted grain) was destined for use in distilling, an industry that expanded significantly during the seventeenth century, but the rest was used for human consumption, or as fodder for animals kept for dairying or meat—and, by extension, also for human consumption. This chapter will next examine in more detail the nature and development of the dependence on and uses of these various grains

FIGURE 1.4 *Calendar Plate for November (Baking Bread)* by Pierre Reymond (France, Limoges, 1561). William Randolph Hearst Collection (48.2.8). Los Angeles County Museum of Art. Public domain.

in the UK and Europe, as well as some of the substitutes proposed at various stages when weather or conflict threatened supply. While grains were consumed in a number of ways, here the focus is largely on bread, one of the most important uses of all the staple grains in production, before touching on the various porridges and gruels they also went in to.

BREAD, THE SYMBOLIC STAPLE FOOD

Bread, the destination for most of the grain grown across continental Europe, the Americas, and Russia, is a food that has been resonant with symbolic power over centuries—arguably millennia—and which simultaneously conjures up a number of different ideas. Although it is no longer necessarily thought of as our staple food, bread continues to be an important component in the Western diet, and in the mid-seventeenth century it still maintained its role at the center of the diet for the vast majority of people. This very place as fundamental building block of all eating makes bread the necessary food that unites people in need, and which implies hard work and its rewards. References to bread can thus also imply a reference to all food, as in the request in the Lord's Prayer to "give us this day our daily bread." This request is only taken by ascetics to mean that people should literally live on bread alone, although in cultures that were still fundamentally Christian, it does imply a suitably humble request only for what is necessary and sufficient. Bread's place in both the old and new testaments of the Christian Bible as emblem of the hard grind of farming, of simplicity, and sharing, and as Christ's body in the Eucharist give it multiple additional layers of meaning. The symbolism of the bread of the last supper, a simple meal shared with Christ's apostles, remained critically important.

The years 1646–51 were referred to in England as "the disastrous years" (Thirsk 1990: 132) of poor weather, poor harvests, and high inflation across large swathes of the country. There was an economic and agricultural crisis. Insufficient domestic grain was being produced whilst the English Civil War had led to trading issues with many former overseas partners, so that it was difficult to import bread grain either quickly enough or at an acceptable price. Concern over shortages was such that in March 1649 the Council of State forbade the export of grain without a license from Parliament or the Council of State. The Essex diaries of the vicar and farmer Ralph Josselin (1616–83) describe repeated early and wet winters during these years; early but cold springs with destructive late frosts; summer flooding and fruit rotting on the trees (Macfarlane 1978: 62–3). The cost of basic commodities reflect these conditions. From January to November 1649 the price of cheese went up from 4d to 4¾d and butter from 6½d to 8d (Thirsk and Cooper 1972: 50–1), increases of 19 and 23 percent, respectively.[1] In the same period, Josselin recorded an increase from 7s 6d to 9s a bushel (a 19 percent increase) in the price of wheat; and an equivalent rise in the price of the cheaper grain, rye. Such an increase in grain prices in particular translated directly into hardship at a time when more than half of a laborer's earnings were spent on basic foods and bread supplied the majority of calories in the staple diet of most people.

In recognition of this reliance on grain and by extension the fundamental place of bread in the nation's diet, the price of wheat bread (including the baker's profit per loaf) had been fixed by law and rigidly enforced according to the Assize of Bread (*Assisa Panis*) since the mid-thirteenth century (not replaced by Bread Acts which regulated loaf size until 1822 and 1836). The Assize determined the size of wheaten loaf one would receive from bakers for one penny (1d), according to the quality of wheat used; so, the price of a loaf was protected, but the size of the loaf would vary (Rubel 2011: 46–7). As grain prices rose so the size of loaves, whether made of the most basic wholemeal or refined white flour, would have reduced while staying at the same price: using the material in the Essex diaries mentioned above, this translates into a variation in loaf size of up to 20 percent.[2] Those eating the finest white breads could always downgrade to the cheaper flour to get more bread for their money (Rubel 2011: 127–8, 40). But, of course, those already at the bottom had no further down to go. The Norman legal system had banned home milling, forcing everyone to have their flour, even in the smallest quantities, ground by the local miller, who was entitled to take 5 percent of it. The grain taken to the mill for grinding was often a mix of wheat and a cheaper grain, usually rye and sometimes barley, which combined to make maslin; the maslin loaf was one of the most commonly eaten breads. Since this type of mixed bread and combined flour was not controlled by the Assize it is quite difficult to monitor market behavior, but it seems clear that the system was abused by bakers and market traders either passing off maslins as pure wheat, or selling at what appeared to be unfair, since unregulated, prices. This hurt most those least able to afford it, according to one reporter pleading for an additional Assize for maslin, since

> the Bread most used to bee put to Sale by Bakers, for the reliefe of the poorer sort, is made of Rye or Masslyn, which is Wheat and Rye or Barley mixed together, in which the poore are daily pinched and wronged.
>
> (Agar 1641: 9)

Bearing in mind that the adult diet depended on at least 1 lb of bread per day per person as the basic building block, and a day-laborer might earn anything from 3 to 12 pence a day, price fluctuations of all kinds made a tangible difference to the available food. Any significant change equated to considerable hardship, especially for larger families, and especially for the already poor.

Although it is clear that grain-based breads were already critical to the diet, agricultural and economic historians have argued that the century leading up to 1750 was the period during which wheat reached its peak as the primary bread grain (Chartres 1990: 201). Its status as the staple grain is directly linked to its status as the most desirable grain for bread-making: every study of bread history establishes the primacy of wheat, and its unique ability to produce the finest, whitest flour and the whitest, lightest bread when sifted finely—white bread being categorically the highest status, most desirable, and most expensive bread throughout this period (Rubel 2011: 40). While some commentators have suggested that the dominance of wheat in this period is overstated, and it is clear that many groups within the population subsisted on cereals other than wheat, even skeptics agree that in the course of the eighteenth century there is evidence of "a growing preference for wheaten bread among all classes" (Wells 1988: 13). As will be discussed in the next section, barley production increased at the same time as wheat production; however, the majority of this production was destined for the growing brewing industry. This meant that barley for domestic consumption commanded higher prices, which made it less available for bread, particularly for poorer households.

This dependence on wheat as the primary grain used in bread, the primary staple food, meant that later in the eighteenth century when population growth, challenges to trade due to European conflicts, and declining domestic harvests led to shortages, proposals for alternatives proliferated. The rice breads proposed by Arthur Young and Thomas Bernard in the 1796 *Annals of Agriculture* (Young 1796: 535–7, 559–60) were typical of many experiments conducted at the time trying to find economical and tasty substitute ingredients for at least part of the wheat in bread. While it is difficult to assess how far such proposals were taken up by the poor people to whom they were directed, it is known that an earlier draft copy of the recipe by Thomas Bernard, a governor of London's Foundling Hospital, is in its archives, suggesting that it may have been attempted in practice and perhaps served at that institution (Bernard n.d. [*c.* 1795]).

The idea of substitution, which was generally proposed by the socially advantaged as a solution to the problems of the socially and economically disadvantaged, was usually accompanied by much moralizing. In 1795 a frustrated Prime Minister Pitt (the Younger, 1759–1806) railed against people's "groundless prejudices" against bread made of mixed barley, rye, and wheat (Hammond and Hammond 1911: 124), and despaired at the southerners' reluctance to eat oats like their neighbors in the north. Even the philanthropist and abolitionist William Wilberforce MP (1759–1833) was infuriated by

FIGURE 1.5 Detail from *Study of Two Women*, chalk drawing by Etienne Jeaurat, *c*. 1755. Purchase, bequest of Helen Hay Whitney, by exchange, 1995. The Metropolitan Museum of Art, New York. Public domain.

and uncomprehending of the stubbornness of the poor when asked to give up white bread for brown. In a speech in February 1800, he lists the recalcitrance and excuses given:

> Even household bread is scarcely ever used: they buy the finest wheaten bread, and declare (what I much doubt), that brown bread disorders their bowels. Bakers do not now make, as they formerly did, bread of unsifted flour: at some farmers' houses, however, it is still made of flour, as it comes from the mill; but this practice is going much into disuse. 20 years ago scarcely any other than brown bread was used.
>
> (Hammond and Hammond 1911: 124)

Given that every household manual of the preceding centuries, and the cookbooks of his own time, consistently highlighted the digestive properties of brown bread versus white, the objections to excessive consumption by the poor of the finer produce is surely as much based on class as simple economics.

WHEAT ACROSS THE WORLD

A focus on bread tends to make a review of wheat a Eurocentric one. But wheat was also, during this period, a fundamentally important grain crop converted into local staple foods in other parts of the world. Wheat is characterized as "the chief grain" across the Middle East, with the exception of places where rice had arrived in previous centuries—southern Iraq and Iran (Perry 2014: 107). In all areas of the region, it was mainly used in breads which were usually flatbreads. Further east, the highly populated grain belt of the Indo-Gangetic plain— stretching from the Indus river in Pakistan to Brahmaputra in Assam and the deltaic lowlands of Bangladesh—produced wheat in vast quantities, as well as rice (Sengupta 2014: 69), making wheat-based flat breads a crucial dietary component for populations in northern India and Pakistan. Equally, the staple grain of northern China was (and remains) wheat, made into noodles, dumplings, and breads (Anderson 2014: 41). While this chapter focuses on western Europe, it is clear that wheat was established as a dominant grain in many other parts of the world during the same timeframe.

BARLEY

Barley is generally thought of as the most important grain for the more basic breads of the seventeenth and early eighteenth century in England, and has been described as "the poor man's usual breadcorn" (Clay 1990: 2, 183). Being one of the staple foods of poverty, barley bread was a useful indicator of hard times in ballads and songs of the period, and frequently associated with humility and penitence (Anon. *c.* 1640). Edward Buckler's pious and profitable meditations of 1640 tell us it is better to live on humble barley bread and pottage made from salt, water, and onion than to get riches by cheating others in trade (Buckler 1640), while numerous sermons exhort greater piety and gratitude by referring to Christ giving thanks for mere barley bread; or recommend godly affliction as a better course than excessive luxury, as shown in Jerome's example of Helarion, who survived for nine years on only six ounces of barley bread per day yet lived his fourscore years; or in the use of barley bread and water by penitential monks. Being well known as poor food, barley bread also formed part of the punishment of condemned prisoners. For example, un-confessed but convicted felons were to be stripped (apart from a loincloth), tied down, and pressed with a weight, then fed on alternate days with three crusts of barley bread and standing water until they died from their injuries, starvation, or hypothermia (Cowell 1651).

This did not mean that barley was only consumed by the condemned and wretched, however, as some people may have preferred it to other grains. Barley bread has a sweet taste, and is a dense, filling food that keeps well. Soaked or dipped in milk or soup it makes a hearty meal. The grain could also be used to make barley milk and barley water, good broth, and a gruel or pudding called frumenty (Thirsk 2007: 219). Later in Herrick's poem "The Hock Cart, or Harvest Home," with which this chapter opened, frumenty is referred to as an "all tempting" component of the harvest supper. Indeed, there are many shades of frumenty (and, as explained below, of gruel) depending on what your wealth determines to be the contents of your larder. The frumenty made with French barley (the best kind, according to Markham) in *The Ladies Cabinet* of 1654 certainly sounds tempting, made somewhat like a rice pudding and enriched with egg yolks and sugar; while a simpler barley pap excludes the egg yolks but is deliciously seasoned with nutmeg, mace, and rosewater (M. B. 1654: 192, 94). Also writing for the more refined table, Robert May included numerous recipes for barley. Although many of them also specify French barley, he recommended a straightforward barley pottage for fasting or Lenten days; a chine (backbone) of mutton cooked in barley broth; and a hot slice of barley bread as a good way to improve the flavor and speed up the making of vinegar. In addition, he recommended it for several "excellent ways for feeding poultry" (May [1660] 1994). Barley was said by Francis Bacon to make the best, healthiest bread for providing solid "aliment of a firm nature" that would ensure the juices of the body were kept beneficially "hard" (Bacon 1638: 266–7). Thomas Cogan, hoping to promote the health of students, was less enthusiastic, saying barley made the worst bread, but nonetheless affirmed it provided the best malt for ale or beer (Cogan 1636: 29–30). These early-seventeenth-century works remained in print as respected volumes for an extended period, and the recipes proposed and ideas about diet expressed remained current into the following century.

RYE AND MASLIN

As outlined above, in many parts of England as well as across Europe wheat and rye were often combined to make maslin. Although the two grains may have been planted in separate fields, according to Thirsk it is also likely that in places the two grains were planted as a mixture and harvested together. While the term might refer to other combinations of grains with wheat, maslin was almost always made with a mix of rye and wheat, in varying proportions, and was the most commonly used bread flour. Thirsk pointed out that numerous different wheats were sown in various parts of the country, and the early ripening cultivars (such as flaxen, pole-eared, and square-eared white) were commonly sown with rye for maslin, as they matured at the same time (Thirsk 1990: 22–3). Although usually sown in the autumn, these cultivars were also sown in spring for summer wheat.

Rye was thought of as simple, honest, godly food, as invoked by John Donne (1572–1631) in a sermon preached at the wedding of his daughter, when he suggested that for the man who is at peace with God, "His Rye-bread is Manna, and his Beefe is Quailes" (Donne 1649: 370). Thomas Moffett (1553–1604), however, took a dimmer view of its common status, saying that although it might cure hemorrhoids, rye was "as

FIGURE 1.6 *Rye (Secale sp.): entire flowering plant with separate flower, fruit and seed.* Colored engraving after F. von Scheidl, 1770. Wellcome Collection. Attribution 4.0 International (CC BY 4.0).

unwholesome for indifferent stomachs, as it agreeth with strong bodies and labourious persons" (Moffett and Bennet 1655: 239). For him, only laborers had the constitutions for tough victuals like rye or brown breads (Moffett and Bennet 1655: 31–2). Country wet-nurses were included in this laboring group, and he blamed their habit of eating over-leavened rye bread (as well as hard cheese and muddy or immature ale) for the illnesses contracted by their infant charges (Moffett and Bennet 1655: 122–3). He considered barley to be just as bad, and maslin "or Munckcorn-bread, made of Rye and Wheate together, is esteemed better or worse, accordingly as it is mingled more with this or that grain"—in other words, more wheat is good; more rye is bad (Moffett and Bennet 1655: 239). Despite these prejudices, though, it is clear that rye was widely used. During the English Civil War, members of the Scottish army reported purchasing a 5 lb loaf of rye bread while in desperate straits in Newcastle, complaining about the outrageous price of 12 pence (Army of Scotland 1640: 9), and a report of provisions in Hull in 1642 includeed fourteen thousand quarters of rye compared to only five hundred of wheat (Anon. 1642: 6). This was grain for the mass of people.

OATS

Although wheat and barley were the most important grains in the more populated areas of England, oats were consumed in larger quantities in northern England as well as in Scotland and Ireland (Wells 1988: 13). Most often consumed as porridges—discussed in more detail in the section which follows—or "hasty puddings" (where boiling water or milk is poured over oatmeal for an instant porridge, as opposed to cooking the grains in the hot liquid for a period of time over the fire), oats were also incorporated into breads, and provided important filler, texture and flavor in sausages, puddings, and haggises (Balic 2013: 84).

GRUEL

Contemporary ideas about gruel have been shaped by the powerful writing of Charles Dickens (1812–70), whose depiction of the thin, watery, starvation rations fed to the unfortunate workhouse orphans in *Oliver Twist* (1839) continues to resonate as the very definition of this foodstuff. But gruel once was regular fare for people in every social class, and thus, like the people, not all gruels were created equal: the porridge or gruel in your bowl varied widely according to your social situation. It is known from the Commission Office's reports (Poor Law Commissioners 1836: 64–6) that the parish-provided gruels, like the ones that Dickens described, were usually truly meager dishes of meal cooked with water, in a proportion of about 1:10 of meal to liquid—more "thick water" than porridge—and served with bread. However, in domestic situations, and even in institutions with more funding, like London's Foundling Hospital, the gruel was different: oatmeal cooked with whole milk, it was a richer and more nutritious breakfast porridge. For the Foundling children, the basic recipe and quantity served did not change throughout the eighteenth and nineteenth centuries: an allowance of half an ounce of oatmeal and a quarter of a pint of milk per child made a tastier breakfast dish, also served with bread (Foundling Hospital 1814).

So, gruels were not only provision for poor and meager meals. Nor were they restricted to oatmeal, or to breakfast. For the middle classes, oatmeal and barley gruels

FIGURE 1.7 *A sick child grimaces as he takes his medication and gruel.* Line engraving by W. Holl, 1838, after W.H. Hunt. Wellcome Collection.

appear in cookbooks as a smooth, digestible food for the sickroom, and as rich supper dishes. Hannah Glasse's *Art of Cookery Made Plain & Easy* contains a section for invalids, which includes several gruels such as her "Water Gruel" made of oatmeal and water enriched with butter, seasoned with salt and white pepper, and passed through a sieve to ensure the smoothest texture (1747: 120). Barley gruels were also commonly found in cookery books as dishes to include in a fast dinner. Glasse's "Plumb-Porridge, or Barley-Gruel" was a mix of equal quantities of barley and dried currants and raisins, flavored with mace, sweetened, and finished with half a pint of white wine (1747: 79). The "Barley gruel" in Mary Kettilby's *Collection Of above Three Hundred Receipts in Cookery, Physick, and Surgery* was recommended as "a wholesome spoon-meat for suppers" (1734: 62). A basic gruel of water and pearl barley is enlivened by dried fruit in the form of currants (dried grapes), enriched with egg yolks and cream, and flavored with white wine, lemon rind, and brown sugar. This is surely more likely to be the kind of gruel the fussy, health-conscious Mr Woodhouse in Jane Austen's *Emma* (1815) had in mind as a late-night restorative rather than the miserable liquid of the workhouse.

RICE

For large parts of the world, rice was the dominant grain. As mentioned above, Asian rice (*Oryza sativa*) was a critical crop in the Indo-Gangetic plain and, whereas wheat was the key crop in northern China, diet and agriculture in southern China were based on rice (Anderson 2014: 41). During the Qing Dynasty which began in 1644 (and lasted until the end of Imperial China in 1911) the population of China more than quadrupled to four hundred million. Rice was a crucial factor in supporting this population as it grew, and in contrast to the industrialization taking place in the global North, in China a focus on small farmers, often at the expense of larger landlords, resulted in a process of "agricultural involution" which meant that while the largely rural population devoted additional labor to improving land management and species development, and produced surpluses, this did not lead to an increase in individual wealth (Anderson 2014: 48). For a more detailed discussion of Qing agriculture than can be managed in this chapter, see Will and Wong (1991) and Hu (2005).

Another species of rice—*Oryza glaberrima*—was the crucial grain in West Africa, used in composed dishes and served with soupy green stews and okra dishes (Harris 2014: 99). Judith Carney emphasizes the historic importance of rice in the West African diet, noting that even today "a meal is not considered complete unless served with rice" (Carney 2001: 8). In *Black Rice* (2001), Carney describes in detail the transfer of African rice-growing techniques and culture (along with the rice itself) by enslaved people transported to the Americas, demonstrating the critical technical and economic contribution made by those individuals to the system that enslaved them and which devalued their knowledge whilst appropriating it. As a result of this transfer process, rice became one of the critical commodity crops for Empire-building and intercontinental trade between the north American colonies and Europe. This trade allowed rice to become established as an inexpensive ingredient for use in British puddings and other dishes (including savory dishes), as well as its use as a substitute ingredient in bread at times of high wheat prices. Rice was also being grown in parts of southern Europe, notably northern Italy and southern France, both for export and as an important component in local diets.

FIGURE 1.8 *Spring in the Rice Fields*, wood block print by Katsushika Hokusai, *c.* 1800. H.O. Havemeyer Collection, bequest of Mrs. H.O. Havemeyer, 1929. The Metropolitan Museum of Art, New York. Public domain.

VEGETABLES AND PULSES

In Britain, besides grains, the seventeenth-century Diggers and their contemporaries were also growing vegetables: beans, peas, and root vegetables such as carrots, parsnips, and turnips, the latter possibly for fodder for cows (Firth 1894: 210–11). The potato was introduced much later, and was being promoted as a staple field crop in many European countries from the 1750s onwards. Roots and pulses, along with greens—both cultivated, such as cabbages, or foraged, such as nettles (*Urtica dioica*)—provided the basis of the diet for the vast majority of people. Improvements in agriculture and the advent of new plant cultivars, especially from Dutch breeders, as well as the development of market gardening techniques practiced across Europe and brought into the UK by both French and Dutch Protestant refugees, were bringing more vegetables into greater production, and thence into the general diet. It is clear from reports of street sellers' and market stall-holders' wares, as well as the dishes proposed in cookbooks of the mid to late eighteenth century, that a wide range of market and kitchen garden vegetables were available not only to those with their own gardens but also to the wider purchasing public. Recipes for all kinds of leaves, herbs, and fancy vegetables like asparagus, peas, and globe artichokes, as well as for soft fruits and salad stuff, show that there was a wide choice of fine plant foods for those with the means to purchase them; and logically we can assume that those groups would define many of those as their staple foods. However, taking a wider view, it is the field crops, produced in far greater quantities, much less prone to damage in transportation, and with greater keeping qualities, that compose the fundamental building blocks of the diet.

TURNIPS, CARROTS, AND PARSNIPS

Thomas Elyot (1490–1546) first expressed attitudes to root vegetables as food in English in the early sixteenth century, building on the work of Galen. At this point, roots were known and clearly accepted as food: "Referring to roots, Elyot expressed a strong view in favour of turnips, since they were thought to augment the seed of man; parsnips and carrots were also nutritious" (Thirsk 2007: 13). One and a half centuries later, a Digger leader, Gerard Winstanley (1609–76), discussed the production of root crops when referring to gardening as "The second branch of Husbandry" (Winstanley 1652: 69). Supporting the suggestion that the Digger plantations combined farming with gardening, "Gardening" in this definition includes knowledge of "how to plant, graft, and set all sort of fruit trees, and how to order the ground for flowers, Herbs and Roots for pleasure food, or medicinal" (Winstanley 1652: 69). As well as being "pleasure food," garden plants in general and roots in particular are discussed as those that are "good for all bodies, both man and beasts" (Winstanley 1652: 69). Blith included carrots, onions, parsnips, and turnips in his "brief discourse of some choice and more general Garden Fruits" (Blith 1652: 259).

Carrots had been eaten in Europe for several centuries. The Greeks and Romans had known them (originating in the Middle East), and Galen had designated carrots as edible although, along with most root vegetables, he considered them to be a "base" food. Although Galen suggested that carrots were not particularly nutritious or digestible, and not recommended in excess, he had to admit they had some value: "they hold less nutriment than the turnip; but they are clearly warming and also are obviously a little aromatic" (Powell 2003: 112). Yellow and red or purple carrots seem to have entered Europe via Spain in the eleventh century, moving into other parts of Europe and reaching England in the thirteenth century. At first, they were viewed as a famine food, and an early French book of proverbs had illustrations that shows men eating grass and carrots next to the epithets "Hunger makes people resourceful" and "Necessity is the mother of invention," but gradually they became accepted as excellent fodder and human food. By the middle of the seventeenth century the orange form most familiar to us today had been developed, probably in the Netherlands, and was recorded in physic gardens in Oxford in 1665 and Edinburgh in 1683.

Charles "Turnip" Townshend (1674–1738) is frequently credited with having introduced the turnip into mass cultivation in the 1730s, inducing Adam Smith (1723–90) to refer in *The Wealth of Nations* (1776) to the switch from the spade to the plow as a driving force for the raising of vegetables—especially turnips (Albala 2003: XI, 33)—as higher-volume field crops instead of mere kitchen garden produce. In fact, Dutch immigrants had introduced turnips in the late sixteenth century (Chartres et al. 1990: ch. 11; Thick 1990). The use of turnips in crop rotation and both as winter animal fodder and food for the poor had already been considered, written about, and implemented in the seventeenth century. Adolphus Speed (fl. 1652–9), a protégé of Hartlib's, devoted a chapter of his *Adam out of Eden* to turnips, "detailing how he advised a gentleman to 'Devonshire' his light soil, grow a crop of turnips as green manure and then put the land down to grass" (Speed 1659: 18–29).[3] Speed mainly focused on turnips' many uses as animal fodder, describing it as nourishing food for almost every farm animal as well as for rabbits, hunting dogs, and horses (when made into horsebread), giving two good crops a year that could be sown immediately after ground was broken and worth, he claimed, £30 per acre (today something between £4,100 and £5,700).[4] Of course, better food for animals also produced more dung to improve soil fertility, a virtuous circle. As well as inducing cows to give more milk, when made into bread in 50 percent proportion with regular meal, Speed found that turnips made a whiter (and thus more desirable) bread that stayed moister and cut better than bread made with meal alone (Speed 1659: 24–5). Despite attributing it to the accumulation of "crude" humor, even Galen (Aelius Galenus, 129– c. 216 CE) had written about turnip with grudging approval: "when boiled in water I should be surprised if it were less nourishing than any plant of the same group. People prepare it in a variety of ways, even to the extent of storing it in brine and vinegar so as to be able to use it for the whole year" (Powell 2003: 110). Even fresh, turnips store well in a cool place, so were a practical choice for human and other animal consumption (Albala 2003: 33).

There is evidence of both turnips and carrots in their red, yellow, and purple forms as an established garden vegetable early in the seventeenth century along the Thames Valley (Kerridge 1968: 177). Thirsk, too, discussed early-seventeenth-century growing of carrots, parsnips, turnips, and other vegetables for the table in the common fields in Fulham near London, in rotation with wheat, with some of the carrot crop used to feed poultry (Thirsk 1990: 29). Besides being fed to the animals, carrots were cooked into soups or puddings (Albala 2003: 33). In late-sixteenth-century works by Richard Gardiner and Thomas Cogan, carrots were highly recommended for all classes, especially as nourishing food for the poor in hard times. In their view they were excellent added to pottage, as well as worthy of being served in place of meat, boiled and buttered, like parsnips (Thick 1998: 21–2). The Diggers said that they would eat their vegetables with pride: "we have peace in our hearts, and quiet rejoycing in our work, and filled with sweet content, though we have but a dish of roots and bread for our food" (Everard et al. 1649: 18). In these root crops, working people had foods that they ate themselves, and used for cattle fodder; and that are also known for breaking up the ground, an important benefit to the farmer.

POTATOES

Often referred to as the fuel that drove the industrial development of the nineteenth century, potatoes were less important than other root crops through most of our period. However, it was during our time that they began to be established. This was in part due to the need to discover new, cheap, and reliable sources of food for growing populations, and also as a solution to the recurring problems of the mid- to late eighteenth century with grain crop yields. An extended period of colder weather affected grain harvests, and it became clear that the potato was productive in these cooler conditions. In France, Antoine-Augustin Parmentier's (1737–1813) chemical study of the potato, published in 1773 with the approval of King Louis XVI, marked the beginning of a fashion for potatoes at court. After the French Revolution, the potato was promoted as the ideal republican food in Madame Mérigot's *La Cuisinière républicaine*, a short recipe book devoted entirely to the potato and containing thirty-one recipes for every course, including potato bread (Mérigot 1794: 36–7) and a sweet "Gâteau économique."

GREENS, BEANS, AND PEAS

Contrary to popular belief, greens have always been a crucial element in the European diet, whether foraged greens supplementing the diet of the poor or perfect leaves from the kitchen and herb gardens of the wealthier classes. As the cities grew, so did the demand

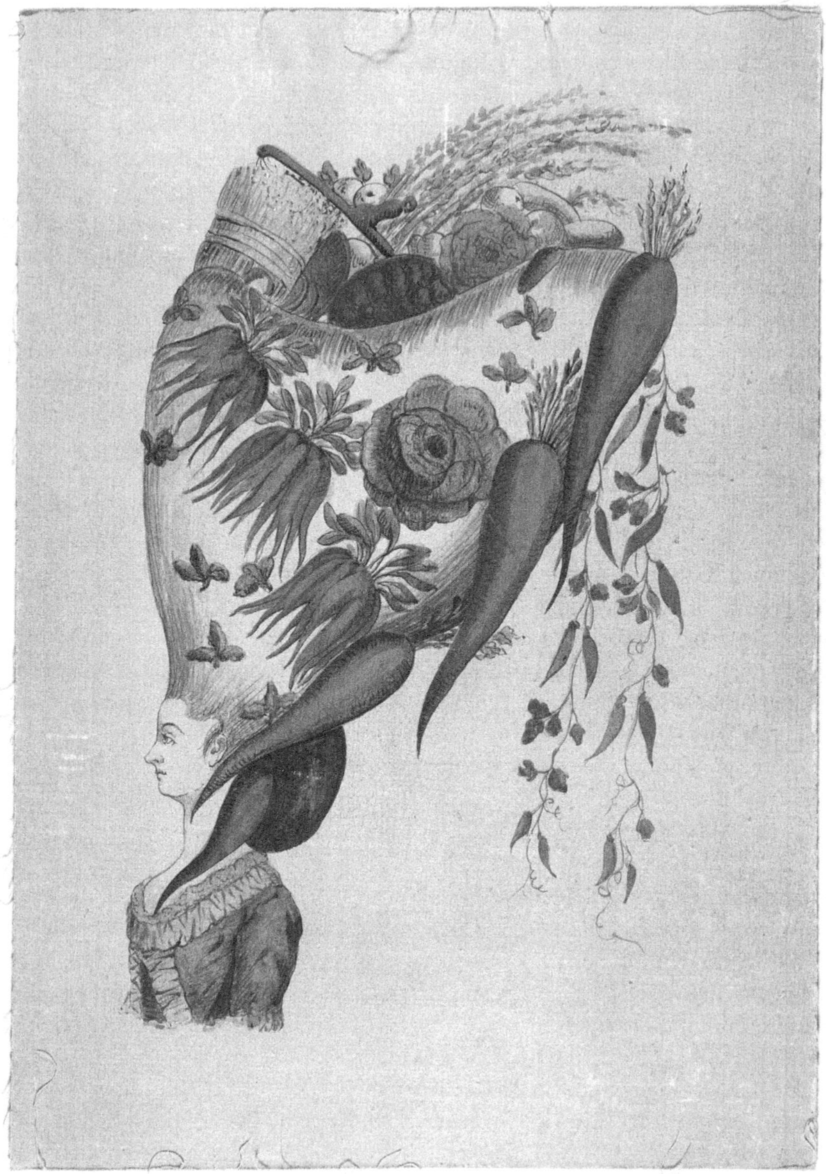

FIGURE 1.9 Anon., *Fantastic Hairdress with Fruit and Vegetable Motif*, eighteenth century. Watercolor on canvas. Alfred W. Hoyt Collection, bequest of Rosina H. Hoppin, 1965. The Metropolitan Museum of Art, New York. Public domain.

for green vegetables. In and around London the market gardening area grew from 10,000 to 110,000 acres between 1660 and 1721, reflecting a growth in production of cabbage and other leaves, salad stuff, fresh peas, and beans as well as other more specialized crops such as asparagus, celery, melons, mushrooms, cucumbers, and radishes which could not be transported long distances so well (Thirsk 1990: 237, 238) and thus needed to be grown closer to market. Greens in the form of cabbages, spinach, and salad vegetables were widely used, the latter especially by the growing middle classes (Thirsk 1990: 235). A "respectable" household in the late seventeenth century served meat "besieged with five or six heaps of cabbage, carrots, turnips, or some other herbs or roots," while a foreign visitor to London in 1748 also found that his hosts ate butcher's meat accompanied by "boiled roots or cucumbers, lettuce, salad and spinach, and ... green beans and peas when they were in season" (Thirsk 1990: 235; 239). A glance at any cookbook of the period shows the core importance of green vegetables and salad leaves throughout the period.

Green beans in the genus *Phaseolus* had been "discovered" in the Americas and introduced to Europe in the late fifteenth century, but even a century or two later these green or kidney beans were not the beans referred to as field crops. Beans from the Americas tended to be known in the period mainly as kitchen garden plants, and although their consumption increased over the period, as such they were less commonly available. For solid sustenance, particularly for the poorer sort of people, the long-established fava or broad beans (*Vicia faba*) formed an important part of the diet. Although in the early humoral tradition "Only labourers were thought to have stomachs strong enough to digest" these dried and cooked beans (Albala 2003: 27), it is actually those that were most commonly grown and certainly more commonly eaten, whether rehydrated in a bean soup or mixed with other ingredients into a more varied stew. From the point of view of husbandry, these legumes, like peas, have the added advantage of being good for the soil, as they fix atmospheric nitrogen into nitrates.

Peas, dried and made into "pease pottage," were extremely commonly grown and consumed in the period. The mealier varieties of field pea, used dried, were even lower on the social scale than beans, but nonetheless formed an intrinsic part of most people's diet. In the post-Civil War cookery book of Robert May (1588–1664), *The Accomplisht Cook* of 1660, there are recipes for pease pottage and various boiled grain dishes known over several centuries such as frumenty (a porridge of cracked wheat boiled in almond or cow's milk for a sweet version or meat broth for savory, sometimes enriched with egg yolks or cream) and beer-based caudles (a sweetened, spiced, drinkable gruel of wheat or oat flour, occasionally enriched with egg) (May [1660] 1994). Pea flour was used in bread, in emergencies or times of shortage, as a substitute for part of the grain flour (Rubel 2011: 123–5). As well as the mealy peas, green garden peas seem to have been relatively widely grown to be eaten fresh (Thirsk 2007: 171–3). Dutch horticulturalists had already developed the snap-pea—peas with edible pods—in the early seventeenth century. According to Joan Thirsk sugar peas, then described as *Pisum saccharatum* (Parkinson 1629: 523), were being grown in quantity by Jonathan Swift's grandfather Thomas in the Wye Valley. Many botanical writers devoted attention to the different cultivars of pea, including sugar peas, though it is unclear whether these were today's petit pois (*Pisum sativum* var. *saccharatum*) or snow peas (*Pisum sativum* var. *macrocarpon*) with edible pods. John Parkinson (1567–1650) even assigned the different cultivars to different parts of the country, emphasizing that on chalky soils, near the Thames, the "First wild pease" are grown (Parkinson 1640: 1058).

Dried pulses were crucial sources of nutrition across much of the world. Food historian Charles Perry noted that in this period chickpeas (*Cicer arietinum*), lentils (*Lens culinaris*), and favas remained major sources of protein across the Middle East (2014: 107). They were equally important in India and Pakistan where the largely vegetarian diet was as much a matter of affordability as religion (Sengupta 2014: 68). In western

FIGURE 1.10 Beans and Peas from John Parkinson, *Paradisi in Sole Paradisus Terrestris* (London: Printed by Hvmfrey Lownes and Robert Yovng, at the signe of the Starre on Bread-street hill, 1629), 523. Research Library, The Getty Research Institute (archive.org).

Europe, dried beans and peas were used in many dishes including hearty soups. One such recipe, Hessian Soup, featured widely in late-eighteenth- and early-nineteenth-century British cookery books such as Mrs Rundell (1824: 138–9), as well as in newspaper articles and manuscript books. An economical dish made of dried split peas and vegetables, thickened with flour, Hessian Soup bears a striking resemblance to Rumford's Soup, famous throughout Europe in the early 1800s. Count Rumford (1753–1814) is widely known for his various technical achievements such as redesigning the domestic fireplace and developing new and efficient ovens and kitchen stoves. A founder of the Society for Bettering the Condition and Increasing the Comforts of the Poor, Rumford redesigned the Foundling Hospital kitchen in the late eighteenth century. He developed Rumford Soup when he was working in Bavaria in the 1790s, as a nutritious yet inexpensive means of feeding the poor in workhouses and soup kitchens. Rumford served on the loyalist side in the American War of Independence, where the British deployed thousands of mercenary troops from Hesse in German, and it is tempting to speculate that Rumford's soup owes at least part of its origin to the Hessian soldiers' rations during the American Revolution. The practicalities on grounds of transport, ease of cookery and cost of ingredients certainly make both soups a practical solution for army rations as well as mass feeding of the poor.

CONCLUSION

The approach to food and its production and supply underwent significant change across the world, as the urgent necessity of feeding large, often poor, urban populations emerged as a real governmental challenge. Some of the solutions put forward in Britain, Europe, and North America set the trend for the more fully developed industrialized period that followed. Yet, within this timespan, it could be said that the basic food supply model underpinning Herrick's poem moves into accelerated decline. The idea that for most people, most of the time, what you eat is what you either grow yourself, or is grown close to where you live, breaks down as industrialization expands, cities grow, and access to commons decreases. The logic of the market economy—developing and taking hold in the period—requires that goods, including food, move to the places they are most desired and therefore command the highest prices. This, in part, is what gives rise to logical oddities like the promotion to the poor of imported rice, never grown within the British Isles, as a plentiful and cheap substitute for more expensive locally grown cereal grains like wheat and barley in bread and puddings. What all of this makes clear is that when it comes to the staple foods upon which everyone in a culture depends, what we depend on is plants.

CHAPTER TWO

Plants as Luxury Foods

GARRITT VAN DYK

This chapter focuses on edible luxury plants, and foodstuffs derived from them, within a European historical context. Food choices changed dramatically in the seventeenth and eighteenth centuries, as trade and exploration redefined the known universe of food and drink across Europe and in the American colonies. Some expensive foods, such as sugar and spices, lost their exclusive status when colonization and increased maritime trade made these comestibles more affordable to a wider audience as "small luxuries." The same trade, however, also brought new plants in the form of foodstuffs and beverages which were unusual and costly.

Novelty and exoticism displaced expense as a measure of social status during this period. Consumption of rare comestibles sourced through a merchant was later overshadowed by the status associated with the domestic cultivation of foreign foodstuffs in the gardens of European noblemen and upper-middle class consumers. Tropical foods, plants grown out of season, and plants manipulated by skilled horticulturists to produce novel results all required specialized knowledge, cutting-edge technology, and the financial resources necessary to acquire the same. This combination of wealth, technology, and refined taste was beyond the means of many, but within the reach of monarchs and affluent consumers, who turned botanical gardens into centers for horticultural experimentation as they vied for the distinction of cultivating exotic foodstuffs in cold climates or out of season.

While the chapter takes a European perspective, it acknowledges here the plant source of foods included in Chinese imperial cuisine during the Qing Dynasty (1644–1911). The elaborate banquets and personal tastes of Emperor Qianlong (1711–99) had considerable impact on the food customs of China, although some of his habits were too expensive to extend beyond the palace. One of those practices was Qianlong's first "snack" of the day—a bowl of crystallized sugar prepared with birds' nests—served before his first audience. Qianlong favored eating at banquets to celebrate special festivals or honor visiting dignitaries. These included many vegetable dishes, and fruits that were served between long pauses in banquets. Bean curd was also prominent and considered to be a tonic food that was good for the health. Tea was the main non-alcoholic beverage consumed. The number of courses was extravagant, with as many as sixteen cold dishes and fifty main dishes; yet food was less important than the many rituals that came before, during, and after the meal (Newman 2008).

In comparison with Europe, the food choices of China were little impacted by the expansion of trade during the period. As Qianlong famously responded to the gifts from George III presented during the Embassy of 1792, "I set no value on objects strange or ingenious, and have no use for your country's manufactures."[1] While this was primarily a reaction to objects—such as Wedgwood pottery, a Herschel telescope, clocks, and watches—it extended to a lack of interest in European plant products to be used in luxury culinary practices.

The examination in this chapter of plants as luxury foods during the seventeenth and eighteenth centuries considers how the rare comestibles of the Old World gave way to exotic comestibles imported from the New World. The development of novel techniques for the propagation of tropical plants in Europe not only transformed the concept of conspicuous consumption to include production in the process, but also permitted the move of species from one part of the globe to another, redrawing the borders of botanical imperialism. To trace the contours of this journey, the first part of this chapter provides some context for the patterns of consumption for luxury plants in the period, followed by an introduction to the most widely consumed luxury foods. It goes on to discuss how the status of imported comestibles declined and a shift towards domestic propagation of exotic edibles became a marker of wealth and luxury in Europe, and an important development for imperialism and colonization.

THE SPICE TRADE AND FRENCH CULINARY HEGEMONY

During the second half of the seventeenth century, increased exploration and expanded trade routes altered the food choices of Europe, impacting the availability of plants used as luxury foods. As consumers responded to the fluidity of meaning and significance attached to specific products, demand for novelty and innovation in metropolitan centers outstripped the supply of new goods. Consumption of the latest imports functioned as a measure of social distinction, as both goods and ideas circulated through local and global trade routes. Attempts by the State to control domestic consumption of imported goods through sumptuary prohibitions in Europe, however, failed to regulate expenditure and consumption to preserve social distinctions.

In a dramatic change of culinary direction around 1600, French chefs began to reject foreign spices, refined cooking techniques, and focused on the intrinsic qualities of the more subtle flavors of herbs and vegetables found within France. There is no definitive documentation indicating what specifically caused French chefs to make these shifts. The most persuasive theory links the turn away from the use of exotic spices to declining prices and competing imports as a result of direct trade. Lower prices eroded the perception of spice as a marker of elite status. This occurred throughout Europe, but exotic spice use continued in English and other Continental cuisines. While cheaper prices would have altered the perceived exclusivity of spice, it does not explain an abrupt rejection en masse of spices in French cookery. Unlike anywhere else in Europe, and especially in contrast to England, the social differentiation previously conveyed by spice was replaced by culinary refinement in France.

As French cuisine became the new luxury, the emphasis on imported spices was supplanted by the domestic production of a French system of delicate eating, which not only served to glorify the nobles who employed the chefs, but also became an export transmitted through cookbooks. The widespread publication of cookbooks, which began with *Le Cuisinier françois* in 1651, enabled this dramatic, and unilateral, move by French chefs away from the homogenous court cuisine which emphasized the use of exotic flavors. While pepper (*Piper nigrum*), cloves (*Syzygium aromaticum*), and nutmeg

(*Myristica fragrans*) endured in minute quantities in French cuisine, other spices such as grains of paradise (*Aframomum melegueta*), galangal (*Alpinia galanga*), and cardamom (*Elettaria cardamomum*) disappeared.

In the preceding five centuries, the demand for spices imported from Asia could aptly be described as a mania. Contemporary historians such as Wolfgang Schivelbusch, author of *Tastes of Paradise: A Social History of Spices* (1992), and Paul Freedman, author of *Out of the East: Spices and the Medieval Imagination* (2008), have compared the economic and political importance of spices in the medieval period to the essential modern commodity—oil. Spices, however, were more expensive with a very high value-to-weight ratio, much prized by shipping companies for the profitable nature of this trade. Beyond their economic value, imported spices were a major feature of court culture, before European trading in the early seventeenth century made them available to a wider market at lower prices. Cloves, mace and nutmeg (both from *Myristica fragrans*), and cinnamon (*Cinnamomum verum*) were luxury items: they were highly sought after for their value as markers of status and exoticism in food and drink at the banquet table; they were in demand for their purported medicinal values, especially in the absence of viable alternatives; and they were required for their dietetic role in balancing the Galenic system of humors.

The Dutch East India Company (Vereenigde Oostindische Compagnie; VOC) had achieved control of the trade in tropical spices—pepper, nutmeg, mace, and cinnamon—sourced from the Moluccan islands in Indonesia beginning in 1602. Initial profits were astronomically high, returning 400 percent on the cost of the voyage, but declining prices led the VOC to establish a ruthless monopoly to limit the supply to the European market. As punishment for violating an agreement to supply mace and nutmeg exclusively to the Dutch, the VOC massacred, tortured, and enslaved almost fifteen thousand inhabitants of the Banda islands in 1621. Despite these extreme measures to ensure a monopoly, competition with the English East India Company resulted in a drop in prices. Using the most expensive of the spices, cloves, as an indicator of the overall market, prices in Surat (Gujarat, India) dropped by 50 percent in the period from 1630–50. Declining profit margins and high costs led the VOC to search for other high-value plants which would yield profits without the expense of warfare (Hochstrasser 2005: 174–5).

PLANTS IMPORTED AS LUXURY BEVERAGES

European exploration introduced exotic new beverages derived from plants and sourced from the Americas and Asia: chocolate (1544), tea (1610), and coffee (1615) (Ukers 1935: 1:23). Sugar complemented the consumption of all these beverages, which rose in popularity from the middle of the seventeenth century. European consumption of sugar increased dramatically in this period, after English and French colonies in the Caribbean established sugarcane plantations based on the exploitation of slave labor. At the end of the eighteenth century, French colonies produced 37.7 percent of the sugar exported from the American tropics, second only to Britain's 38.8 percent. St. Domingue (now Haiti) was the largest single sugar-producer in the Caribbean, with an output almost twice that of Jamaica (Sheridan 1974: 101). These two islands were the most financially successful European plantations, and their production supplied the growing demand for sugar in North America and Europe. The increased availability of sugar changed its status from an exclusive commodity available only to the wealthy to a small luxury reaching more modest consumers (Deerr 1950).

As early as the end of the seventeenth century, sugar was used on a daily basis by a quarter of the English population (Shammas 1990: 81–2). Continental consumption also grew throughout this same period, mainly in urban centers, but habitual use was limited to middle-class consumers. Even though some beverages, such as tea, were not always sweetened in their country of origin, European and American consumers frequently added sugar.

CHOCOLATE

The earliest in Europe of these beverages, chocolate (*Theobroma cacao*), was introduced to Spain in 1544 by nobles of the Q'eqchi' (Kekchi) Maya of Central America, who were brought from Guatemala by Dominican friars to visit the court of Prince Philip. The delegation presented exotic gifts from the New World to the royal court: quetzal feathers, clay vessels, gourds, chilis, maize, beans, sarsaparilla, and prepared drinking chocolate in clay vessels. While the young prince, later Philip II, recorded observations of the Maya, he did not comment on the chocolate. Nonetheless, the drink must have made enough of an impression to encourage the first shipment of cocoa beans to Spain from Veracruz, Mexico in 1585. Despite the unusual combination of spices and the unusual heat of chilis, chocolate became a popular drink, both in Spanish colonies and in Spain. Beyond the unfamiliar flavors of the beverage, chocolate also had significant meaning for Mesoamericans, linking humans with the gods. Chocolate was offered in tribute to deities, and consumed in rituals related to birth, death, marriage, and festivals. The scientific name given to the species and genus by Linnaeus in 1753, *Theobroma cacao,* reflects this connection with the divine. The combination of the Greek words for god (*theos*) and (*broma*) food were combined with the Mayan word *cacao*, and became the taxonomic designation: cacao, food of the gods (Coe and Coe 1996).

Spanish colonists' interest in chocolate was originally focused on its medicinal properties and pharmacological effects. As the plant was transported from the New World to Europe, this trend continued, despite the established reception of chocolate as a social beverage in New Spain. The first recipe for chocolate published in Spain appeared in 1631 in a medical treatise published in Madrid: *Curioso tratado de la naturaleza y calidad del chocolate* by Antonio Colmenero de Ledesma, a physician. The preparation of the "king of chocolates" called for "one hundred cocoa beans, two chillies, a handful of anise seed and two of vanilla, two drams of cinnamon, one dozen almonds and the same amount of hazelnuts, half a pound of white sugar and enough annatto [*Bixa orellana*] to give some color." All ingredients were boiled together, and then frothed in a jug or chocolate pot with a *molinillo,* a carved wooden stick with a ribbed sphere at its base. The *molinillo* was placed in the jug and held between the palms of the hands, which when rubbed back and forth frothed the contents of the vessel. Other recipes called for native ingredients such as "ear flower" (*Cymbopetalum penduliforum*), called "xochinacaztli" in Nahuatl or "orejuela" in Spanish. The unavailability of indigenous ingredients did not prevent Europeans from attempting to recreate the original flavors with readily available spices (Grivetti and Shapiro 2009).

This effort to simulate the traditional ingredients and methods has been seen by Marcy Norton (2006) as an internalization, albeit unintentional, of Mesoamerican aesthetics. The lack of mediation in consumption of a symbolic native food was a significant departure from the ideology of imperialism. From the viewpoint of the European colonizers, they brought technology, civilization, and salvation to the indigenous population. In this instance, however, the habitual adoption of a religious comestible, even when stripped of its sacred significance, represented an inversion of the pattern of conquest.

Consumption of drinking chocolate was so popular in the early modern Iberian world that it became a point of contention with the Catholic Church. The popularity of chocolate was noted in a treatise of José de Acosta (*c.* 1540–1600), *The Naturall and Morall Historie of the East and West Indies* [*Historia natural y moral de las Indias*, 1590; with the English translation of 1604 attributed to Edward Grimeston]: "The Spaniards, both men and women, that are accustomed to the country, are very greedy of this chocolate ... they make diverse sortes of it, some hote, some colde, and some temperate, and put therein much of that chili; yea they make paste thereof, the which they say is good for the stomacke, and against the catarre" (Acosta 1604: 271). The question of whether drinking chocolate was allowed during periods of fasting or abstinence established by the Catholic Church was addressed by Tomás Hurtado (1570–1649) in a 1645 treatise on chocolate and tobacco, *Chocolate y Tabaco Ayuno Eclesiastico y Natural*. He concluded that a water-based recipe did not constitute a "food," but recipes that called for milk or eggs were not permitted. Despite this favorable determination, the bishop of Chiapa (later Chuapas, Mexico) found that the consumption of chocolate during Mass was an affront punishable by excommunication. Upper-class female parishioners had been drinking chocolate during Mass, brought into church by their maids, disrupting the service and interrupting sermons and prayer. When the bishop prohibited chocolate, these ladies began to attend Mass at cloisters outside the city, until the bishop prohibited city residents from worshipping anywhere but in the cathedral. The women refused to comply with this order and remained at home instead. The Bishop, however, fell violently ill and was covered in painful lesions for a week before he died. The cause of his illness was claimed to be a poisoned cup of chocolate, giving rise to a famous proverb: "Beware the chocolate of Chiapa" (Gage 1648).

This story was related to English readers by a Dominican friar, Thomas Gage (*c.* 1603–56) in his treatise on the New World, *The English American: A New Survey of the West Indies* (1648), along with several variations of chocolate preparation for different medical ailments. Chocolate was consumed at all but the lowest social levels in New Spain and the West Indies, and had also been adopted by the upper classes in Flanders, Italy, and Spain—but had not made the same impression in England. Chocolate was first advertised for sale in London 1657 in the *Publick Adviser* (No. 4, June 9–16): "In Bishopsgate Street in Queen's-head Alley at a Frenchman's house is an excellent West India drink, called CHOCOLATE, to be sold, where you may have it ready made at any time, and also unmade at reasonable rates." Further encouragement was found in Henry Stubbe's *The Indian nectar, or, a discourse on Chocolata* of 1662. Stubbe (1632–76), physician to Charles II, drew on a wealth of botanical and medical treatises from the New World and Spain to produce a comprehensive evaluation of the nutritional value of chocolate. Through the mediation of the English author, many of the Spanish texts were received by Stubbe's audience as an endorsement of chocolate as a healthy food, suitable for the constitution of an Englishman. Dr. Stubbe served chocolate to Charles II, confirming that it was a drink suitable for the monarch. He listed various possible additions to the chocolate preparation, such as eggs, milk, and sack (fortified white wine from Spain). Stubbe claimed that chocolate, on its own, was a healthy food but the ingredients added in more elaborate recipes might be harmful. Even without the more exotic elaborations, chocolate was an expensive beverage for wealthier consumers and nobles.

The luxurious nature of the beverage, its exotic origins, and relative expense led to associations with decadent behavior and patterns of consumption. Chocolate houses opened in London beginning in 1657 with an unnamed establishment in Bishopsgate Street. This was followed by the eponymous Cocoa Tree, Ozinda's and Mrs White's Chocolate House (1693) in Mayfair. The last of these, founded by Italian immigrant Francis White (Francesco Bianco), became notorious for debauchery and gambling, being described by Jonathan Swift as "the common Rendezvous of infamous Sharpers [= swindlers] and noble Cullies [= dupes] ... the Bane of half the *English Nobility*" [Swift *c.* 1720: 127]; it survives as White's, a private gentlemen's club in St. James'.

The luxury status of chocolate changed as the plant was introduced to European colonial plantations or was found growing wild in newly acquired territory. Early imports of *Theobroma cacao* (native in South America) from Mesoamerica in the seventeenth century were supplemented by the Spanish exploitation of wild cacao growing in Ecuador, and high-quality *criollo* cacao grown in Venezuela. By the eighteenth century, chocolate production changed again as plantations were established in the West Indies (Guadeloupe, Martinique, St. Domingue, Jamaica, and Trinidad), and mechanized processes for grinding cacao were developed in France. With increased crop production and the advent of hydraulic and steam-powered grinding, the price of processed chocolate fell and became more widely available. Despite the advances in mechanization, and considerable demand, the technology necessary to transform cacao into chocolate bars was not developed until late in the nineteenth century.

TEA

By contrast with chocolate's early arrival in Europe, tea (*Camellia sinensis*) was first introduced by the Dutch East India Company (VOC) in 1610, imported from their base on the Japanese island of Hirado. The VOC was granted trading privileges on the island in 1611, but conflict with Portuguese priests led to the expulsion of all European inhabitants. Unable to establish a significant trading presence the VOC turned to China as the source for tea (Ukers 1935: 1:29). The growing trend for tea consumption in The Netherlands can be seen in the instructions sent in 1637 from the VOC directors in Amsterdam to the governor-general of Batavia (modern Jakarta, Indonesia): "As tea begins to come into use by some of the people, we expect some jars of Chinese, as well as Japanese tea with each ship" (outgoing letter book of the VOC, Heren XVII to Anthony van Diemen, January 2, 1637; quoted in Schlegel 1900: 469). Despite the regular shipments, the quantity of tea imported was still quite small, and prices were very high. A pound of imported tea in 1660 cost nearly sixty times the daily wage, making it a luxury that only wealthy consumers could afford (Martin 1832: 16). The first public auction of a commercial quantity of tea in Amsterdam was not until 1667 (Parmentier 1996: 78). By 1673 the VOC allegedly employed a Dutch physician, Dr. Cornelis Bontekoe (Cornelis Dekker 1640: 7–85), to publish a tract *Tractat van het excellenste kruyd thee* (Treatise on Tea the Most Excellent Herb) proclaiming the curative properties of tea. This popular treatise was reprinted in three editions and was influential in the promotion of tea as a beverage in the Netherlands, and more generally in Europe (Schweikardt 2003). While tea was also drunk in Germany, France, Russia, Spain, and Portugal during this period, the habit was most deeply embedded in the fabric of daily life in eighteenth-century Britain.

In contrast to the early adoption of tea in the Netherlands, the English East India Company (EIC) began its pursuit of tea much later in the seventeenth century.

Correspondence between EIC factors in Kyoto and Hirado in 1615 mentions tea, but only for personal consumption. The growing importance of tea to the EIC, however, can be noted from the value placed on the new commodity by the Directors in 1664. They wanted to make a gift to Charles II and requested an appropriate present to be selected from recently arrived ships, laden with tropical foods, precious goods, and foreign animals: "Captain Prowd is directed to inquire aboard the several ships for rare birds, beasts, or other curiosities fit to present to the King" (Sainsbury 1925: 52). The royal gift selected from the multitude of exotic goods was tea. While the Directors originally feared the gift was poorly chosen, the response from Charles II was so positive that a greater quantity was procured and offered the next year.

The choice of tea as a gift from the East India Company to the King of England reflects not only that it was appropriate as a gift for the monarch, but also that it was symbolic of the endeavors of the Company. An entry in the EIC "Despatch Book" in 1667 confirms the first commercial quantity of tea brought into England by the company: "We desire of you to procure and send by these ships … 100 weight of *Tey*" (BL IOR/E/3/87, ff. 65v, 137). This import marks the beginning of the EIC's rise to its position as the biggest importer of Chinese tea, amplifying the importance of the company's monopoly on trade between Asia and Britain. In 1713 the EIC was granted access to Canton (modern Guangzhou), and large quantities of tea were imported from 1717 onward (Erikson 2015: 125).

Once a gift fit for a king, by the end of the eighteenth century tea was available throughout Britain, even to its poorer inhabitants. This did not prevent noble and upper-middle-class consumers from developing modes of consumption to differentiate their tea drinking from the daily draught of the lower classes. Unfermented green teas (singlo, hyson, and gunpowder) were more expensive than fermented black teas (bohea, souchong, pekoe, and congou). Black teas were, on the whole, more popular as they paired better with milk and sugar added to balance bitterness and tannins. They were also less likely to spoil in transit (Nierstrasz 2015: 266). Tea consumed by the poor was often low-quality black tea, sourced from the VOC and smuggled into Britain. Subject to duties of up to 119 percent, tea was a considerable source of customs revenue, leading to widespread smuggling (Nierstrasz 2015: 272). For consumers of limited means, cheaper tea was also available through merchants, but it was often adulterated with tree-leaves dyed to look like the genuine product or contained tea that had spoiled in transit. Tea imported by the VOC often fell into this category, as the company used bamboo tea chests, resin from which spoiled the contents. Lead-lined EIC tea chests were more dependable for safe transport of tea, but were heavier, and therefore more costly to ship (Glamann 1958: 221).

Beyond the quality of the brewed beverage, the ritual of tea consumption was a gendered and socially distinctive practice. Outside the home, tea was served to mixed company in tea gardens, or to a largely male clientele in coffee houses. Tea service in the home was overseen by the lady of the house. After dinner, tea was served to female guests while male guests drank alcohol. Tea drinking in polite society was accompanied by expensive tea *equipage*: porcelain (or silver) teapots, porcelain tea bowls and saucers, and silver tongs for handling small pieces of sugar, broken from the large conical loaf with nippers (Tibbles 1994: cat. 148). Until the eighteenth century, the finest porcelain was imported from China. The techniques for reproducing Chinese hard-paste porcelain were not developed in Europe until 1708, in Saxony, Germany. Similar items were also produced for chocolate (and coffee) service but the paraphernalia required for tea was a more important element of social display as part of the ritual of domestic respectability that constituted the British "tea ceremony."

Tea prices fell when the excise duty of 119 percent dropped to 12.5 percent with the passage of the Commutation Act in 1784. The dramatic reduction of the duty made smuggling uneconomical and encouraged even greater imports of tea. Widespread diffusion of tea brought the beverage within the reach of nearly all inhabitants, and only the differentiation achieved through the accompanying rituals and accoutrement retained any status of luxury. The new habits adopted by the lower classes in the last quarter of the eighteenth century were denounced by commentators, including Scottish politician and judge Duncan Forbes:

> when the opening of a trade with the East Indies ... brought the Price of Tea ... so low, that the meanest labouring Man could compass the Purchase of it ... when Sugar, the inseparable Companion of Tea, came to be in possession of the very poorest Housewife, where formerly it had been a great Rarity ... the effects were suddenly and severely felt. (Forbes 1744: 7)

Tea drinking may have promoted sobriety, but the expenditure on tea and sugar by the poor was considered an unnecessary and inappropriate outlay by the poor who drank tea in imitation of the upper classes.

COFFEE

The last of the caffeinated tropical beverages to arrive in Europe was coffee (*Coffea arabica*). Trade at the Red Sea port of Mocha introduced Dutch merchants to coffee, sold and grown in Yemen under tight controls to ensure their monopoly. The sale and export of raw coffee seeds and plants were strictly prohibited to prevent propagation of coffee outside of Yemen. After initial purchases of coffee beans were brought back to the Netherlands in 1616, Dutch interest in the coffee trade increased (Ukers 1922).

Merchants from the English East India Company also imported coffee from Yemen, which coffee was first publicly available at a coffee house in 1650 in Oxford at the Angel (Cowan 2008: 90). In the middle of the seventeenth century, the London coffee house emerged as a public venue characterized by egalitarian male sociability and informal networks of information. These spaces were not remarkable for their refined service of luxury beverages, but for the type of discourse conducted: science, commerce, journalism, politics, and literature. In the years following the interregnum, the coffee house filled a vacuum created by the beginning of the Restoration, fueled by a common desire for novelty, news, and unfettered discourse (Wild 2005: 90). The coffee-house model was so successful that by the beginning of the eighteenth century there were nearly six hundred in London alone.

French merchants imported coffee into France, first in Marseille in 1644. A coffee house opened there in 1671, but the first Parisian café did not appear until ten years later. An enduring example of the seventeenth-century café can be seen in Café Procope, still in operation in Paris. Designed as a fashionable interior with elaborate décor, the café was a space for sociability and display, an alternative to the salon or the royal court. These new public places helped popularize the consumption of coffee in England and France and encouraged merchants to increase their imports.

Dutch traders were not content with importing coffee and tried to circumvent the controls on export of coffee plants from Yemen in 1658. An initial experiment to cultivate coffee in Sri Lanka from plants left behind a century earlier by Arab merchants failed (Ukers 1922: 6). In 1696 transplanted coffee trees were smuggled from Malabar, southern India, by the VOC. These plants were propagated from seeds that had been smuggled into India from

Mocha in the sixteenth century by a Sufi pilgrim returning from Mecca. The Malabar plants were introduced to Java but did not survive a succession of earthquakes and floods. Finally, in 1699, more cuttings from Malabar were successfully established in Java, producing enough coffee beans and seedlings for shipment to Amsterdam in 1706. Correspondence from Governor-General Jo[h]an van Hoorn (1653–1711) in Jakarta to the VOC directors in 1707 relayed his enthusiasm, and surprise, at the progress of the plants (Glamann 1958: 207). These seedlings from the Jakarta gardens were installed in hothouses with other exotic plants in the Amsterdam Hortus Medicus, later Botanicus, founded in 1638.

By 1714, the Hortus hothouse yielded a coffee bush 5 feet tall that was presented to Louis XIV by the burgomaster of Amsterdam (Ukers 1922: 6). The plant was quickly moved from the royal Château de Marly, near Versailles, to the Jardin du Roi (founded in 1626 and now Jardin des plantes) in central Paris, where it was carefully tended under the instructions of the naturalist, Antoine de Jussieu (1686–1758). This single plant was the progenitor of colonial coffee production not only in Martinique and the Île Bourbon, now Réunion, but also South America, Central America, and Mexico (Spary 2012: ch. 3). Many of the leading producers of coffee today owe a distant debt to the coffee plants that came from Java, via an Amsterdam hothouse. French planters were late in the race to grow coffee in France's colonial territories in the early eighteenth century, compared to the initiatives of Dutch colonists who began planting coffee in Java almost twenty years before (Samper and Fernando 2003). The Dutch East India Company had discovered decades earlier in the spice trade that the cultivation of self-perpetuating agricultural resources in new colonial territories was more effective than importing or manufacturing in achieving economic self-sufficiency for the colonies.

In contrast to French colonists' efforts, English planters in the Atlantic never made any serious attempts to grow coffee on a large scale. Instead, early colonists preferred to use colonial acreage for a crop complementary to English consumption of tea—sugarcane. In each case the patterns of consumption mapped to the trading strength of the nation-state and, where possible, luxury plants were grown in colonial territories rather than being imported. The challenge to grow tropical luxury plants in colonies and in Europe was taken up by the large trading companies, state-sponsored scientists, and by wealthy individuals with experimental hothouses.

GROWING PINEAPPLES

The importance of this development in technology and techniques to grow exotics extended beyond the introduction of novelty foods and creating ornamental gardens. The ability to grow tropical plants in Europe was also linked, on the one hand, to the re-creation of an earthly Eden, and on the other hand, to aspirations of botanical imperialism—the relocation of acclimatized species to colonial possessions for profit and power. The possibility of Eden on earth is portrayed in the 1629 title page of *Paradisi in Sole Paradisus Terrestris* by John Parkinson (1567–1650). The woodcut by the Swiss artist Christopher Switzer shows all the plants gathered by Adam and Eve that would re-create, as the Latin title (also a pun on Parkinson's name) suggests, an Earthly Paradise (see Figure 2.1). The fruits of empire are also central to the frontispiece in *Historische Beschryvinghe van Amsterdam* (1663) by Olfert Dapper (1636–89), wherein the personification of Amsterdam as a female allegorical figure receives tributes from four continents, the world genuflecting before her. In the background, a ship at sea reminds the viewer of the source of this bounty. Such tributes were not merely allegorical—travelers, merchants, and visiting dignitaries hoping to garner favor with monarchs brought delicacies from the four corners of the earth, delicacies which were then presented to guests at courtly gatherings as symbols of extra-territorial influence.

The fruit that symbolized this combination of exoticism, novelty, and colonial conquest was the pineapple (*Ananas comosus*). In his diary, the English landscape architect and gardener John Evelyn (1620–1706) speculated that the first pineapple to arrive in England was given as a present to Oliver Cromwell in 1657 (Evelyn 2000: 293). Unfortunately, he did not provide any details about who gave it to the Lord Protector, or if Cromwell ever tasted the fruit. Perhaps a pineapple, referred to here as a "King-Pine," was a gift more appropriate for a king, because Evelyn provided a full description of the reception of the French ambassador, Charles Colbert, Marquis de Croissy, at the court of Charles II in 1668:

> I saw the magnificent Entrie of the Fr: Ambassador *Colbert* received in the Banqueting house ... Standing by his Majestie at dinner in the Presence, There was of that rare fruite called the *King-Pine*, growing in *Barbados* & W. Indies; the first of them I had ever seen; His Majestie having cut it up, was pleasd to give me a piece off his owne plate to tast of, but in my opinion it falls short of those ravishing varieties of deliciousnesse, describ'd in *Cap: Liggons* history & others; but possibly it might be, (& certainly was) much impaired in coming so farr.
>
> (Evelyn 2000: 513–14)

The building described, the banqueting house at Whitehall, was a separate space where guests retired to be served the final course of a formal meal (Strong 2003: 200–1). "Banqueting stuffe" included various forms of confection: marzipan, biscuits, candied spices, poached fruit in sugar syrup, candied flowers, marmalade, fruit pastes, and wafer cakes, transformed into pyramids or presented in geometric arrangements (Markham 1986: 190). Pride of place in Evelyn's description, however, was given to the pineapple, which was carved ceremoniously by the King himself. Both France and England had colonial possessions in the West Indies, so the pineapple was not unknown to the French ambassador, but it was still exotic enough that it would have served as a symbolic reminder to the French ambassador in 1668 of England's continued strength in the arena of global trade, despite her recent defeat in the Second Anglo-Dutch War (Beauman 2005: 45–8).

The difficulty of conveying this experience was underlined by English thinker, John Locke (1632–1704), in his essay *On Human Understanding*, published in 1690. He used the pineapple as the centerpiece of his argument that true knowledge can only be based on experience:

> if you doubt this, see whether you can by words give anyone who has never tasted pineapple an idea of the taste of that fruit. He may approach a grasp of it by being told of its resemblance to other tastes of which he already has the ideas in his memory, imprinted there by things he has taken into his mouth; but merely raising up in him other simple ideas that will still be very different from the true taste of the pineapple.
>
> (Locke 1990: 7)

Moreover, Locke argued in *An Essay Concerning Human Understanding* that "nobody gets the relish of a pineapple till he goes to the Indies, where it is, and tastes it," despite the fact that Locke did not travel to the Indies himself (Locke 1959: 2:328).

In response to the desire for first-hand experience, and the perishability of exotic foods, botanists attempted to grow exotic plants in Europe in artificial environments. The

FIGURE 2.1 Title plate to John Parkinson's *Paradisi in Sole Paradisus Terrestris* by Christopher Switzer in 1629 (https://www.metmuseum.org/art/collection/search/355128). Courtesy of Gift of Estate of Marie L. Russell, 1946, The Metropolitan Museum of Art.

manuscript for John Evelyn's *Elysium Britannicum* reflects the effort to raise pineapples, noting in the margin next to an illustration of the fruit: "But his Majesty having had divers of these sent him over & even ripening after they were here in is garden at St James, we are not altogether to disparage propagation of them" (Evelyn 2001: 414). His note mentioned the ripening of the fruit and differentiated this from its propagation, meaning that the pineapples in the royal garden were grown from established but immature plants. This was no mean feat, considering that in a tropical environment, a flowering pineapple plant will take six months to produce fully ripened fruit. Raising a pineapple from a cutting in seventeenth-century England would have taken three years and been difficult without the ability to maintain the correct temperature and exposure to sunlight. This had to be accomplished without exposure to smoke from an open fire. Pineapples can thrive on limited water and in partial shade, but they cannot tolerate frost. A painting from *c.* 1675–80, *Charles II Presented with a Pineapple* (see Figure 2.2), attributed to Hendrick Danckerts (*c.* 1625–*c.* 1685), shows the presentation of a pineapple to the King by the royal gardener, John Rose (*c.* 1621–77). Earlier claims that the painting depicted the first pineapple "grown" in England are unfounded, since such an achievement would have merited more attention and would have been thoroughly documented.

FIGURE 2.2 *Charles II Presented with a Pineapple*, attributed to Hendrick Danckerts (*c.* 1625–*c.* 1685). The Royal Collection Trust, London. Photo by Historica Graphica Collection/Heritage Images/Getty Images.

The first pineapple claimed as grown in the Netherlands, and possibly Europe, in the garden of Agneta Block (1629–1704) in 1685 attracted significant interest (Beauman 2005: 60).[2] Block, originally married to a rich silk merchant, was widowed at forty and used her inheritance to buy a country estate on the river Vecht bear Utrecht to realize her vision for an experimental botanical garden. She transformed the established garden with the introduction of exotic species that she had commissioned the VOC to collect for her and raised them in greenhouses on her estate (Kearney 2012: 72). Her achievement was commemorated by a silver medal (see Figure 2.3): one side of the coin bears her profile, while on the other, Block is depicted as *Flora Batava* with a cornucopia in her left hand, and a tulip in her right hand. She is flanked by a pineapple, and a cactus, both plants growing in pots. The inscription at the top of the coin bears the name of her garden estate, *Vyverhof,* which is shown in the background. At the bottom of the coin, the inscription in Latin reads "*fert arsque laborque quod natura negat*" (art and labor produce what nature cannot), celebrating Block's ability to control nature through her gardeners' technical skill. The inclusion of the cactus and the tulip in the composition emphasizes both the exoticism of her endeavors and the national identity of Dutch botanists. Her

FIGURE 2.3 Agneta Block (1629–1704), First Person in the Netherlands to Cultivate a Pineapple, Jan Boskam, 1700, striking (metalworking). Rijksmuseum, Amsterdam. Public domain.

achievements are also shown in a painting (*c.* 1694) by Jan (Joannis) Weenix (1640–1719), a family portrait, with a pineapple plant with mature fruit growing in the left-hand corner (see Figure 2.4). The composition leads the viewer on a diagonal from the left to the centre, where the subject is flanked by her two step-children and her second husband. In the background, the gardens of Vijverhof are depicted, linking Block and her family to the cultivation of exotic plants, further emphasized by showing the entire pineapple plant, rather than just the harvested fruit.

The cutting from which Block's pineapple was raised was most likely from specimens brought to the botanical gardens of Leiden and Amsterdam from the Dutch colony of Surinam in 1680 by a ship's captain (Wijnands 1988). Surinam was administered as a commercial entity by the Society of Surinam, owned by three shareholders: the Dutch West India Company (the WIC), the City of Amsterdam, and the relations of the founding colonial governor, the van Aerssen van Sommelsdij[c]k family (Ness and Cope 2016: 616). Directors of the Society were also directors of the Dutch East India Company (the VOC), and the botanical garden of Amsterdam (Plotkin 1992: 78). The overlap in governance of the Amsterdam Hortus with the directors of the WIC, VOC, and Dutch colonies provided an extended network for the circulation of foreign plant specimens destined for public and private gardens in the Netherlands. Established in 1590, the Hortus Academicus in Leiden received specimens from the VOC at the request of the original curator, the French botanist, Charles de l'Écluse (Carolus Clusius; 1526–1609),

FIGURE 2.4 Jan Weenix, *Family portrait of Sybrand de Flines (1623–97), Agnes Block (1629–1704) and two children in their country seat Vijverhof*. Amsterdam Museum, Amsterdam. Public domain.

making it the dominant center for botany in the Netherlands (Wijnands 1988: 62). The Amsterdam Hortus Medicus was replaced by the Hortus Botanicus in 1682. Joan Huydecoper van Maarsseveen (1625–1704), a founding commissioner of the new Hortus, was a burgomaster and VOC director. Through his connections, the Amsterdam collection grew to rival that of the Leiden Hortus.

The exchange between colonies trading with the WIC provided further opportunities to expand the number of species represented in the Horti and, in turn, the gardens of private horticulturists, like Agneta Block. In 1675, William III of Orange requested that the VOC directors supply him with an annual delivery of plants, animals, and "curiosities" from the East Indies ("Catalogue/Catalogus" 1988: 292). While such gifts helped the VOC secure the continued support of the Stadtholder, company directors were also concerned about their monopoly on exotic goods being undermined by the proliferation of species outside of their control.

As indicated above, the acquisition of viable specimens was only part of the challenge in propagating the first pineapples. Sunlight and a constant temperature are necessary for the growth of the plants, unlike, say, citrus trees which are less sensitive to cold, and are dormant during winter. Bitter orange trees had been successfully cultivated in northern Europe since the middle of the sixteenth century, sheltered from the cold winters in basic shed-like shelters, later called orangeries, with rudimentary heating from charcoal braziers and narrow windows providing a small amount of light (Grant 2013: 7). The first record of an attempt to grow a pineapple in the Netherlands, by doctor, botanist, and naturalist Bernardus Paludanus (Berend ten Broeke 1550–1603) in 1592 failed because the delicate plants were too sensitive to the cold. Paludanus noted this as an annotation on an entry in Jan Huyghen van Linschoten's *Itinerario* of 1596: "I have had some of the Slips here in my garden, that were brought mee out of Brasilia, but our colde countrey could not brooke them" (Linschoten 1885: 17–18). New technology was required to re-create the environment in which to grow tropical plants. Stove heating was introduced in orangeries, but fluctuations in temperature and the dryness of the air were unsuitable for delicate exotics. For better temperature control, narrow structures were designed as lean-tos, against the exterior wall of an orangery. Alternative heating methods were introduced to provide constant radiant heat, using vertical ducts in the wall, or horizontal ones under the floor. Fireplaces located next to, or under, the hothouse fed these ducted systems (Sikkens-De Zwaan 2002: 212–14). Tropical hothouses were shown on plans for the Amsterdam Hortus from 1682 and were already in use in private gardens in the Netherlands, such as Block's Vijverhof, by 1685 (Wijnands 1988: 76).

With a stable ambient temperature, the remaining challenge for reliable propagation was a constant soil temperature. John Evelyn described a hotbed with fermenting manure to provide warmth, and a glass-covered frame which he specified as being "for such plants as require the utmost degree of this artificiall heate to raise and entertaine them" (Evelyn 2001: 109). Manure was not a perfect solution, as the heat did not last, and the manure needed to be replaced on a regular basis, with the added risk of disturbing the plants in the process. Finely ground oak bark, left over from leather tanning, was used by silk merchant and naturalist Pieter de la Court van der Voort (1664–1739) in his hothouses at Allemansgeest, his estate in Voorschoten near Leiden, to provide a lasting source of bottom heat for growing pineapples ("Catalogue/Catalogus" 1988: 281). De la Court used this method to great success, reliably producing full-sized fruit on a regular basis. His methods were widely copied in Britain and on the Continent, later being published in his 1737 text *Byzondere aenmerkingen over het aenleggen van pragtige en gemeene landhuizen, lusthoven, plantagien* ... (Special Remarks about the Design of Country Houses, Pleasure Gardens, Plantations ...). Translated into German and French, the text provided detailed instructions on the propagation of tropical plants in the cold climate of northern Europe, the use of tanner's bark, and illustrations of the tropical hothouses used at Allemansgeest (Beauman 2005: 61–3).

The influence of Dutch methods, and the speed with which ideas circulated, can be seen in the results achieved in England by Henry Telende, the Dutch gardener to the wealthy Sir Matthew [Mattijs] Decker (1679–1749), a director of the EIC born in the Netherlands. De la Court's methods had been outlined by English naturalist, Richard Bradley (1688–1732), a Fellow of the Royal Society, in his *A general treatise of husbandry and gardening,* published in 1721, some sixteen years before de la Court's book was released. Following de la Court's methods "which Mr. Telende ... has rendered so easy and intelligible, that I hope to see the *Ananas* flourish for the future in many of our English gardens, to the Honour of the Artist, and the Satisfaction and Pleasure of those that can afford to eat them" (Bradley 1721: 209).

Bradley anticipated the dissemination of the techniques for successful propagation, but underestimated how far the fashion would spread, or how long it would last. The mania for growing pineapples transmitted from the Netherlands to Britain spread across Europe to Russia. During the reign (1762–96) of Catherine the Great, pineapples were grown in most houses of the Russian nobility. By the first half of the nineteenth century the number of pineapples grown in the *Tsarskoye selo* hothouses reached a thousand in pots. In 1801, a pineapple hothouse, later known as a pinery, was built in at the Russian Imperial residence, Pavlosk, the residence for Tsar Paul I's widow, Maria Fedorovna (1759–1828). Designed in the English Palladian style by the Scottish architect Charles Cameron (1745–1812), hothouse peaches and pineapples were ripened in the pinery despite the freezing Russian winters.[3] One hundred years after Richard Bradley commented on how pineapples would soon be enjoyed by a growing audience, another British commentator was still amazed by the ability to cultivate the plant:

> The quantity of pines and grapes grown in the neighbourhood of Petersburg, is indeed an astonishing feature of its horticulture. Pines, grapes, and peaches being grown so as to ripen in August and September, enjoy, in these months, abundance of sun, and nearly equal in flavour those grown in England or Holland ...
>
> (Loudon 1822: 56)

Even more remarkable is Loudon's frame of reference, which compared the Russian pineapples to those of England and Holland, and not to those in the tropics. The imagery of the pineapple, and its evocation of both an earthly Eden and the reach of Empire, was found in European architecture, interior decoration, and in early shop advertisements. From the western towers of Christopher Wren's design for St. Paul's Cathedral, to political cartoons, the wallpaper of Parliament, trade cards, and the extraordinary pineapple that adorned the walled orchard and summerhouse at the estate of the Earl of Dunmore in Scotland, the transplantation of the pineapple had not only occurred physically in the pineries of Europe, but also in the imagination of its inhabitants.

CONCLUSION

The technology that replicated the environment of the tropics was not limited to growing exotic plants for consumption in Europe but was also used to study those that might be propagated in overseas territories. Removed from their original growing areas in the

Americas or Asia, plants were transported to botanical gardens in Europe to determine their viability as crops that could be grown domestically or sent out to the colonies. Eager to reproduce the success achieved by the cultivation of sugarcane in the Americas, European trading companies searched for new plants which would yield profits through a combination of high prices and low weight.

Without the controlled climate of the tropical greenhouse, developed in large part through the experimentation of amateur botanists, the coffee seedlings imported from Java would not have been able to survive in northern Europe. Despite the global connections and financial resources of the VOC and WIC, and their intertwined relationship as both collectors for, and sponsors of the Horti at Leiden and Amsterdam, it is somewhat surprising that the most significant achievements in propagation were not made by state-sponsored institutions, but by private individuals. In the case of pineapples, the gardeners of amateurs like de la Court and Agneta Block accomplished what royal gardeners and professional botanists were unable to achieve.

Indeed, the contributions of private individuals have been underestimated by garden and food historians alike. The lack of institutional hierarchy and procedure allowed for greater freedom in methodology and experimentation outside the boundaries of accepted academic protocols and the exigencies of commercial viability. Both the circulation of ideas within their networks of amateurs, as well as the exchange of this knowledge with the wider scientific community, contributed directly to the propagation of exotic plants and the development and consumption of luxury food items, not only in Europe, but increasingly in the colonies.

An enduring example of this can be seen in enormous experimental vegetable gardens at Monticello in Virginia, USA (see Figure 2.5), developed by the gardeners, including slaves, of Thomas Jefferson (1743–1826). Monticello grew fruits and vegetables from

FIGURE 2.5 Thomas Jefferson's experimental garden at Monticello. Image by AgnosticPreachersKid/CC BY-SA 4.0.

across Europe and the Americas, sourced through international seed exchanges and exploration. Jefferson's scientific field trials determined which of three hundred kinds of some ninety species were best suited to the Virginia soil. After serving as president, Jefferson declared himself freed from the shackles of government and dedicated himself to the pursuit of gardening in service to the new nation. He counted his successes in the garden as more important than his accomplishments in government. Rather than citing the Declaration of Independence as the pinnacle of his achievements, Jefferson declared that "The greatest service which can be rendered any country is to add a useful plant to its culture" (Hatch 2012: 4).

CHAPTER THREE

Trade and Exploration

SARAH EASTERBY-SMITH

On September 20, 1726, officials in the French port town of Nantes received a royal decree from the king, Louis XV. Nantais merchant ships, the decree explained, were now required to collect and bring back seeds and even live plants from the overseas colonies that they visited. These seeds and plants would be cultivated first in the Apothecaries' Garden in Nantes. Once they had been "naturalized" in the French climate, cuttings would be transferred to the royal botanical garden in Paris, the Jardin du Roi. As the decree itself noted, many of France's port towns—not least Nantes—already boasted gardens that grew exotic plants imported from overseas. The twin intentions behind the decree were thus to formalize and expand an existing custom of importing exotic plants and to ensure that the new specimens would ultimately reach the Jardin du Roi, France's premier institution for botanical research (ADLA C632, Ordonnance 1726).

Such an instruction would have been a source of great delight to scholars in France concerned with inventorying the natural world and with applying knowledge about plants to improve medicine, agriculture, and industry. The decree would guarantee not only the regular arrival of exotic specimens, but also would ensure that the plants were passed immediately to individuals with the necessary expertise for their cultivation and examination. In the three years that followed, the mayor of Nantes, Gérard Mellier (1674–1729), sent reports to Paris in which he itemized the specimens that he had received, listing them by name and indicating where they had come from and who had conveyed them (AMN DD50 and DD51). Consignments arrived irregularly, often with long intervals between one parcel and the next. Each normally amounted to between six and twelve items, packaged up in boxes or barrels and accompanied with a list of the contents and an *attestation* from the ships' officers. This latter document described the plants' conditions when they were laded and testified to the care taken to nurture them while on the vessel. The tender packages were watered regularly with fresh water, and were usually placed in the stern of the ship—the area least subject to its pitch and roll while at sea, and where they would be sheltered from pernicious salt water (AMN DD51, Folder 68). Such precautions should have ensured the plants' survival as they were carried across the oceans, and thus would contribute to increasing the quantity and range of exotic specimens cultivated in France.

In spite of the royal directive, however, only a very small proportion of the total number of merchant ships actually fulfilled the terms of the decree and imported living plants to France.[1] Furthermore, those that did attempt to bring back new plants generally found that the majority died during the voyage or—in the case of seeds—that they would not germinate on arrival in France (AMN DD51). The scheme was not wholly successful, then, being impeded first by an apparent reluctance to comply with the royal orders, and then by the high number of losses at sea (Parsons and Murphy 2012). Nevertheless, the French government persisted for decades afterwards, reissuing the instructions to its merchant navy to bring back plants. The government's perseverance indicates the value placed on obtaining new plants—and on transferring relevant information about them—by almost all European states in the long eighteenth century. But the existence of such a scheme also calls into question the nature of the support offered for eighteenth-century plant collecting, and its connections to the newly emerging science of botany. Why were the government's directives adhered to so unevenly? Answering this question involves engaging with the multiple forms of value attributed to the collection and study of plants during the Enlightenment.

Trade and exploration were mutually reinforcing elements in the cultural history of plants, especially in the eighteenth century. Seeking new markets, merchants had over centuries developed long-distance trading connections that came to link disparate lands and peoples within global networks. European explorers initially followed those trade routes, and the plants they gathered were normally sent back to Europe on merchant ships. The connection between trade and exploration extends further, however, beyond the practical question of transferring plants globally, to that of European colonization. Seeking new commodities, and new locations in which these might be produced, processed, or sold, merchants—and especially mercantile syndicates such as the European East India Companies—aimed not only to discover new lands but also to gain control over these places and the peoples who lived there (Batsaki et al. 2016; Kumar 2015; Iliffe 2008; Schiebinger and Swan 2005; Stewart 2008). Native flora (and fauna) might constitute either the raw materials from which commodities could be fashioned, or essential nourishment for settlers. Plants could be turned into objects for further trade, or might furnish cures for the many ailments that afflicted travelers (ADLA C632, Note 1786; ADLA C632, Drogues n.d.; Cook 2007). The search for knowledge about plants, which included information about their properties, about how they grew, and about how different species might be related to one other, motivated and facilitated travel, trade and, ultimately, empire. To trade and exploration, then, must be added two further structural elements key to eighteenth-century plant collecting: developments within the natural sciences, and the expansive imperial ambitions of eighteenth-century European states.

This chapter examines the relationships between these four overlapping themes within Enlightenment plant collecting: trade, exploration, botanical science, and the state. It discusses the problems involved in transporting plants across the globe, the significance of contemporary debates about the usefulness of natural resources, especially exotic plants, and the range of individuals involved in gathering and growing new specimens. It will become clear that multiple economies emerged around the possession of such specimens: the values attributed to plants ranged from the economic and political to the social and symbolic. The contemporary emphasis placed on collecting and studying plants reflected a broader Enlightenment culture of empirical enquiry and useful application of natural knowledge. Thus, while this chapter primarily focuses on examples from eighteenth-century France, it will make comparisons with other parts of Europe, especially Britain. Trade and exploration were certainly key to the cultural history of plants in eighteenth-century Europe, but the ways in which new specimens were acquired, and how they were then used, also developed in response to the emergence of new forms of value attributed to botanical knowledge, and to the shifting priorities and fluctuating involvement of governments and members of civil society.

SCIENCE AND EMPIRE

France mostly obtained new plants via its imperial connections. Broadly speaking, France commanded two kinds of empire through most of the eighteenth century. To the west, the French competed against other European powers, most notably the British and Spanish, in seeking to create what might be described roughly as an empire of settlement. A booming transatlantic trade, primarily based on the economics of slavery, but also on trade in other key commodities including cod and furs, drove French expansion across the Atlantic World, especially in the Caribbean and Canada. The French imperial approach to the Americas was shaped by the desire to access these raw materials, and by competition with other European powers for the possession of land (Delbourgo and Dew 2007; McClellan 2010; Parsons 2017; Schiebinger 2004b). To the east, on the other hand, the French and other European powers focused their efforts on developing mercantile empires that were somewhat less concerned with the formal control of territory. French commercial interests here were largely concentrated on the Indian Ocean, and were normally directed by the Compagnie des Indes Orientales (the French East India Company). Like its other European counterparts and competitors, France established key trading posts across Asia, especially around the Indian subcontinent. It also colonized several Indian Ocean islands, especially Madagascar, the Île de France (Mauritius), and Île Bourbon (Reunion) (Banks 2003; Pluchon 1991; Quin 2000). These islands, especially the Île de France, acted as essential stopping-off points for French ships, allowing crews to recover and restock supplies before continuing their grueling journeys onward across the oceans.

France's eighteenth-century colonial possessions were relatively small in number compared to those of other major European powers such as Britain and Spain. The fortunes of the French empire also shifted significantly over the course of the century. Its Indian trading posts were periodically lost and regained, and its empire as a whole saw a major contraction at the conclusion of the Seven Years War in 1763. The Compagnie des Indes was completely disbanded between 1769 and 1785. But France nevertheless managed to retain its most lucrative colonies. The wealth accumulated from these possessions was most visible in the port towns that lined the Atlantic coast, not least the city of Nantes.

Botany occupied a significant place alongside sciences such as cartography and astronomy in serving European overseas expansion. Over the long eighteenth century, increased travel, either through trade or exploration (or both), meant that the number of exotic plants introduced to Europe also expanded dramatically (Bourguet and Bonneuil 1999; Cooper 2007: ch. 2; Hall 2003: 5–9; Mukerji 2005: 30). This increase was in spite of the high mortality rates experienced during transit. These plants were identified, variously, as economically useful resources, or as beautiful, ornamental specimens; alternatively, they might have no known properties and thus require further investigation. The Nantais example mentioned above underlines the attempts made by the French Crown to coordinate the collection of plants. Recognizing the huge gains to be derived from the collection and study of data gathered overseas and brought back to the metropole, the French state gradually introduced a series of measures designed to increase the certainty that it would receive potentially valuable information. Louis XV's proclamation of 1726, for example, was intended to create a seamless chain stretching from France's global trade networks to its port cities and onward to Paris.

Studying initiatives such as this proclamation, historians James E. McClellan III and François Regourd have depicted the overall structure created within Old Regime France as a "colonial machine," through which new data and information were collected and processed (McClellan and Regourd 2001; McClellan and Regourd 2011). According to this model, science in eighteenth-century France evolved directly under the auspices of the Absolutist state, whose ability to collect information, and then to put it to use in order to harness and exploit its colonial resources, apparently resulted in French science becoming "institutionally and intellectually the strongest of any nation" (McClellan and Regourd 2011: 32). The appealing image created by this historiographical model, of centralized control, uniform direction, and efficient application, carries significant limitations, however. As intimated above, a range of different constituencies in fact supported the collection and study of plants from overseas. Certain participants were more responsive to mercantile incentives than to governmental ones. Thus, at times exploratory plant collecting in the eighteenth century was carried out in accordance with government policy, but in other instances it took place in contexts that were in fact relatively independent from the state.

Examples of these various instances will be reviewed below, starting with a discussion of the development of economic ideas about natural wealth, which shaped French approaches to imperialism and encouraged formal state involvement in plant collection. The roles played by members of civil society, whose activities might play counterpoint to those of the state, will be discussed in the final section of this chapter. What will become clear is that the profit economy to be derived from plants encouraged both state and private patronage of plant collecting, but financial gain and political approbation were not the only forms of value associated with this activity. Social competition over possession of rare specimens, and a symbolic economy derived from the prestige associated with botanical scholarship and horticultural skill, also encouraged the collection of plants. These four mutually coexisting forms of value determined the extent of support offered for plant collecting, and the kind of person willing to invest time, money, and effort into obtaining new specimens.

PLANTS, POLITICS, MONEY

The systematic study of plants' properties, growth, and classification had developed initially between the fifteenth and seventeenth centuries as a branch of medicine. In France, the transition of this body of knowledge into an independent discipline—botany—was tied up with royal aspirations for control over natural resources. As E. C. Spary has shown, the late-seventeenth-century French state established a number of high-profile schemes to support the search and repatriation of drugs considered key to ensuring the health of the nation and its colonial settlers; compared to earlier centuries, the political value placed on medical botany was extraordinary, and this area of enquiry consequently gained a prominent public profile (Spary 2004: 15–20). By the early eighteenth century the Crown had developed or presided over further initiatives that, although focused on deploying botanical knowledge for medical use, further underscored a connection between national prosperity and all forms of applied botanical knowledge. The study of pharmacy (which included *materia medica*) was made compulsory for medical students in 1707 (Spary 2004: 16); the following year a royal directive was issued explaining that the *Premier Médecin* had Paris' Jardin du Roi entirely at his disposition for study and teaching (ADLA C632, Lettres 1708).

The earlier, narrow focus on medicine broadened out to encompass other contexts in which plant knowledge might be applied usefully, especially agriculture. Natural history, and specifically the science of botany, thus emerged in France (and elsewhere in

Enlightenment Europe) as an arena of knowledge that was linked directly to arguments about utility and the exploitation of natural resources or—in other words—political economy. As historians including Londa Schiebinger, E. C. Spary, Margaret Schabas, Fredrik Albritton Jonsson, and Peter Jones have shown, European economic ideas about national wealth were developed through and in relation to debates about the best means of deploying natural resources (Albritton Jonsson 2013; Jones 2016; Schabas 2005; Schiebinger 2004b: 5–6; Spary 2004). The most dominant strands of thought were those collected under the umbrellas of physiocracy, cameralism, and mercantilism. The former, which was largely elaborated by a coterie of French economic thinkers in the mid-eighteenth century, identified agriculture as the sole source of wealth, and thus opposed domestic and overseas trade and industry, perceiving both as "sterile" activities. Physiocracy's non-interest in trade and its limited application means that it will not be discussed at length here (in spite of its firm advocation of the manipulation of the natural world). By contrast, the latter two strands of thought were directly concerned with commerce and industry, and sought in particular to control capital flows by imposing restrictions on the circulation of commodities.

Proponents of cameralism encouraged an isolationist approach, seeking to close borders to exchange in order to keep money circulating solely within the specific national unit. Mercantilists developed a more geographically expansive view, aiming to restrict the flow of bullion within the confines of an oceanic empire rather than within a national unit. Both groups viewed natural resources as a significant source of wealth, and were thus very interested in exploring the potential contributions that botanical knowledge might make to national prosperity (Albritton Jonsson 2013: ch. 2; Jones 2016: ch. 2). Plant collecting and study were upheld as means of improving agriculture and medicine at home and (for mercantilists) in overseas empires.

Cameralists, who included the Swede Carl Linnaeus (1707–78) and his French rival, George-Louis Leclerc, comte de Buffon (1707–88), each hailed by his devotees in eighteenth-century Europe as a "father" of natural history, sought to identify and "improve" native plants and to acclimatize exotic species within national borders. A range of domestic policies emerged that encouraged the surveying of indigenous resources. Linnaeus explored Swedish Lapland, and then sent a succession of his "disciples" back there to continue his work in cataloging Sweden's northern riches. Inspired in part by Linnaeus' work, institutions elsewhere in Europe likewise surveyed domestic resources. Edinburgh Botanic Garden, for example, sent its gardeners out on missions to inventory systematically the hitherto unfamiliar resources in the Scottish Highlands and Hebrides. Across Europe, an increase in the production of county natural histories, local floras, and even statistical surveys responded to the same impulse to enumerate the wealth that already existed within national borders (Albritton Jonsson 2013: 58–9, 63–4; Hodacs 2011: 183–209). The history of eighteenth-century botanical "exploration," then, includes the banausic histories of interior discovery as well as the more exhilarating, and perhaps better-known, chronicles of overseas encounter and exchange.

The alternative to identifying and "improving" native plants was to obtain completely new species that might supplement the indigenous flora. While Linnaeus sought to cultivate exotic species in Sweden, his mercantilist counterparts elsewhere in Europe saw colonies (which were, of course, also known as plantations) as key assets in this regard (Albritton Jonsson 2013: 59–61; Drayton 2000; McClellan and Regourd 2001: 40–4, 50). The British transplanted European species in their colonial possessions—most especially following their settlement of Australia, where the land, especially around

FIGURE 3.1 *Poivrier (Piper nigrum)*. Watercolor by Delahaye, plate I in *Florindie ou Historiephysico-economique des vegetaux de la Torride* (1789). Photo by DEA/G. DAGLI ORTI/ De Agostini via Getty Images.

Port Jackson, proved amazingly hospitable to European migrant flora. Transplantation schemes also involved transferring non-European species, such as spice plants, across the world. The French, for example, spent decades attempting to obtain the "true" nutmeg (*Myristica fragrans*) from the Dutch, ultimately managing to obtain and cultivate it in their

Indian Ocean colonies. They and other European colonial nations likewise experimented with acclimatizing plants in new parts of the world. East Asian plants such as pepper (*Piper nigrum*), mango (*Mangifera indica*), and breadfruit (*Artocarpus altilis*) were sent to grow in West Indian colonies, for example, and sugar (*Saccharum officinarum*), cotton (*Gossypium* spp.), and indigo (*Indigofera* spp.) were cultivated in the French Mascarene Islands (Bret 1995; 1999: 65–89; Drayton 1999; Ly-Tio-Fane 1970; Roberts 2014: 323–7; Spary 2005).

Although broader histories of eighteenth-century Europe tend to characterize nations as conforming to one particular economic doctrine, few (if any) actually saw a national consensus over how wealth might best be created, or over the relationship between trade and wealth. Britain, for example, saw competing schools of thought about the best method of extracting wealth from the natural world emerge. In England, major patrons of natural history such as Joseph Banks developed "neo-mercantilist" schemes that were focused on the exploitation of nature (Gascoigne 1998: ch. 5). Across the border in Scotland, however, the ethos was broadly cameralist. As Fredrik Albritton Jonsson has shown, Scottish improvers adopted—and adapted—aspects of cameralist doctrine to suit their own situation (Albritton Jonsson 2013: 62). In France, the institutions created by the Absolutist state encouraged schemes to collect and process natural knowledge that embraced the whole range of economic ideas about wealth, including economic theories unsympathetic to trade, such as Physiocracy.

Certain French royal institutions played a clear role in supporting exploration and the collection of plants, especially from overseas. In founding the Académie Royale des Sciences in 1666, Louis XIV and Jean-Baptiste Colbert (1619–83) had made explicit the fact that science could be made a tool of royal power. Government patronage of science was channeled almost exclusively via the Académie and other royal institutions, such as Colbert's Observatoire Royal, the Jardin du Roi, and the Académie Royale de Marine (which was founded under Louis XV, in 1752). Funding and support were thus placed in the hands of a select few, and French scholars (especially those who received a *pension* from the Académie Royale des Sciences), had an obligation to serve the King and his ministers (Hahn 1971).[2] Questions of utility occupied a prominent place within the enquiries undertaken by the Academicians (Briggs 1991). The dominance over French science of these scholarly institutions contrasted hugely to other European countries, especially Britain, which did not see such centralization.

However, although the existence of these institutions gives the impression of direction from above, French royal support for plant-collecting schemes normally emerged in response to supplications from individual promoters and could not be expected to continue over the medium and long term. Early in the eighteenth century, for example, botanists at the Jardin du Roi had to petition the King for royal support for botanical exploration, putting forward a case for the value of such activities (Spary 2004: 18–19). In 1726 Louis XV, who was generally sympathetic to entreaties from the Jardin botanists, followed this up with the directive mentioned at the start of this chapter, a directive that was reissued periodically during his reign. Although the Crown's responses to such requests were often positive, royal patronage was neither guaranteed nor, if offered, necessarily sustained. This was the case on both sides of the English Channel. Larry Stewart has recounted the sorry history of Scottish medical student and plant hunter Robert Millar, who was sent by the trustees of Britain's Georgia colony in the 1730s to gather Caribbean specimens suitable for transplantation to Georgia. In spite of the significance of his collections, Millar's patrons were not able to find him an adequate position back in

Britain, and he died shortly after his return (Stewart 2008: 831–2). Millar's tale was not unusual. The French surgeon and botanical collector Joseph de Jussieu (1704–79), who accompanied Charles Marie de La Condamine on a 1730s South American expedition funded by the French crown, was—quite literally—abandoned. Louis XV refused to renew or extend the financial support for the expedition, and his decision meant that Jussieu, who had been separated from the main body of the mission, was stranded in the subcontinent without funds to return home—ultimately for thirty-six years (Safier 2008: 203–24). In spite of their general receptiveness to economic arguments in favor of collecting and studying plants, and despite the general encouragement of global plant-collecting schemes, European governments did not offer consistent, sustained support for botanical exploration. This resulted not only in failure to identify and obtain new natural resources, but also in the loss of botanical collectors themselves.

LIMITATIONS OF STATE STRUCTURE(S)

The image of a state-run "colonial machine" thus presents an overly totalizing model for colonial science, and for plant collecting specifically.[3] Significant qualifications of that model, in addition to those sketched out above, exist on several fronts. These relate, firstly, to the government's ability to compute the volume of information it received; secondly, to the relationship formed between the government, the Compagnie des Indes, and the French colonial botanical gardens; thirdly, to the organization of voyages of exploration; fourthly, to the existence of networks that bypassed metropolitan centers; and, finally, to the role played by civil society in providing the infrastructure and financing to support plant-collecting schemes. The first four of these limitations to the power of the state will be briefly outlined and then, in a separate section, the final point about the significance of civil society will be discussed more extensively.

The first qualification to the "colonial machine" model, quite simply, is that the French state was not able to process the vast quantity of information that it received from its colonies. As Kenneth Banks has shown, the government experienced an "information overload" by the second half of the eighteenth century. The volume of correspondence that the Ministère de la Marine alone received tripled between 1713 and 1763, and the existing bureaucracy was not able to adapt in order to absorb and apply the new information as it arrived (Banks 2005: 19, 26–8). Spain experienced the same problem (Bleichmar 2016: 35–60). The aspiration of European countries to obtain and then control colonial natural knowledge was thus stymied by their own bureaucratic inefficiency.

Secondly, much plant collecting depended on cooperation with joint stock companies. This was especially the case for botanical scholars whose interests were oriented towards the east, where the Compagnie des Indes directed most overseas trade and thus facilitated overseas travel in search of new species. The Compagnie, another of Colbert's creations (founded in 1664), went through several iterations over the course of the eighteenth century, but remained under close royal control throughout (Haudrère 2005). In contrast, the other European companies were directed by merchant governors, rather than by crowns. These included not least the Dutch Verenigde Oostindische Compagnie (VOC) which set the standards for the collection of information about plants, and for the use of that information for colonial control. French circumnavigator Louis Antoine de Bougainville (1729–1811) remarked with astonishment in 1771 that, thanks to its application of natural knowledge, the VOC was "more like a powerful Republic, than a company of Merchants" (Bougainville 1771: 367). Observations such as that made by

Bougainville encouraged a shift in policy towards the systematic deployment of scholars by the East India Companies, and thus of the collection of data (Stewart 2008: 830–3).[4]

In the French case and, arguably, in those of most other European colonies, scholarship of the natural world was undertaken through processes of negotiation between European scholars and numerous other authorities. The latter might comprise local rulers and informants, a European state, and the trading company or other mercantile syndicate (Gasgoigne 2014; Kumar 2015; McAleer 2016; Roberts 2014; Sivasundaram 2010). There are some significant instances where these different authorities collaborated in the interests of furthering the collection and study of plants. In the French case, Crown and Compagnie supported, for example, the establishment of new botanical gardens in France's colonies, including on the Île de France, Île Bourbon, Saint Domingue, Guadaloupe, and Cayenne (McClellan and Regourd 2001: 42). Botanical gardens were consistently located in significant centers for trade and were essential in facilitating further exploration. The gardens acted as "hospitals" where fragile flora (and fauna) might rest before continuing their journeys towards Europe. They were also havens for traveling European naturalists (of any nationality) sent out to explore the wider world (Bravo 2005: 49–65; Bret 1995; Brockway 1979).

Even in the case of botanical gardens, however, cooperation between Crown and Compagnie was not as straightforward as it first appears. The garden on the Île de France, for example, had been founded by the Compagnie des Indes in order to provide medicinal supplies to its ships. Yet the botanists working at this island garden soon expanded upon their original brief, conducting extensive experimentation with the acclimatization of non-native plants (Roberts 2014: 325). As Chandra Mukerji has underlined, the French crown under Louis XIV had been initially reluctant to support the foundation of experimental colonial botanical gardens (rather than medicinal ones), because it was convinced that all new specimens should be brought to France for study. Attitudes shifted greatly from the 1760s, however, as new European ideas about "stewardship" of colonial space gradually gained traction (Mukerji 2005: 27–9, 31–3). The reduction in number of France's overseas territories after the Seven Years War further encouraged a more focused investment in its remaining colonies. By the latter third of the eighteenth century, France, like other European powers, had established a global network of colonial botanical gardens in its trading posts and colonies. These gardens played an essential role in supporting the collection and study of plants. Further, although they were ostensibly under the direction of the state, it will be seen below that the colonial botanical gardens actually enjoyed a relatively high degree of independence.

The eighteenth-century voyages of exploration are probably the most widely remembered instances in which Enlightenment exploration and scientific research came together under Crown direction. While individual European nations had been sending out explorers for centuries, the eighteenth century saw a significant shift in the nature of such exploration, which became explicitly "scientific" in terms of its aims and organization. From the 1760s onwards, royal governments worked with their navies to send out voyages on missions to measure and observe astronomical events such as the Transit of Venus, and to map the unchartered lands around and within the Pacific Ocean. The ostensible intention was to solve major astronomical and thus navigational problems, but these voyages were also equipped with a full complement of competent naturalists, who preserved and packaged up flora, fauna, and ethnographic materials, and with artists trained to portray the strange worlds through which they traveled (Iliffe 2008: 618–45). France led the way with Bougainville's circumnavigation between 1766 and 1769; his mission was shortly followed by that of British Lieutenant James Cook aboard *Endeavour*. The latter brought over a thousand dried specimens of new species of plants to Europe, most of which were donated to the new royal botanical garden at Kew. The botanical dimension of Cook's expedition, however, emerged largely on the initiative of a private individual, Joseph Banks. Banks joined the voyage at his own expense and also provided the personnel trained to observe and collect naturalia (Beaglehole 1969: 512–14; Iliffe 2008: 627). While the value accorded to plant collecting was increasingly brought into state agendas that had been worked out within the political economies of imperialism, the practice of overseas plant collecting continued to be organized through a much looser assortment of private individuals and government officials. This was the case throughout the final third of the eighteenth century.

On arrival at the trading posts and colonies, the travelers on these voyages encountered a range of local experts, including Europeans settled either temporarily or permanently in these imperial spaces, and non-Europeans who were either indigenous or migrants to the same colony. Gathering data was never a question of simply observing the world and objectively recording observations, although European travelers often represented this as such in their written accounts of their travels (Bleichmar 2012; Pratt 1992). The fourth point, then, is that the colonial machine thesis tends to misrepresent the nature of colonial spaces, which constituted centers of collecting—and thus of expertise—in their own right. Each of the four themes outlined earlier—trade, exploration, botany, and the state—featured to a various extent within these locations.

European travelers depended on the connections they forged with the peoples they encountered to obtain new plants. Those who settled in a particular location had a greater potential for creating links with possible informants. Jean Nicolas Céré (1737–1810), the Director of the Jardin du Roi on the Île de France, developed not only an extensive understanding of the local conditions and environment on the Mascarene Islands, but also of the Indian Ocean World at large, thanks to his own networks of correspondents and traveling informants (Easterby-Smith 2016: 19; Ly-Tio-Fane 1970).[5] Historians have conventionally depicted colonies such as the Île de France as satellites from which new knowledge would be transmitted back to European centers (McClellan and Regourd 2001; Mukerji 2005; Paskvan 1971; Stewart 2008: 834). Céré and his counterparts, however, not only developed schemes to collect and study the flora and fauna in their region but also participated in information networks that did not necessarily consider Europe as their central point of focus, or that even bypassed Europe completely. The 1726 proclamation discussed at the start of the chapter sought to create a direct channel along which information might flow from the colonies to Nantes and then on to Paris, where data could be pooled, analyzed, and put to use. The center-periphery relationship that existed between Paris and Nantes, however, was not replicated reliably at the larger scale of the empire. Increasing evidence is emerging that shows that such connections were modified, or broke down entirely, further afield (Brixius 2015, 2017; McAleer 2016; Roberts 2014; Safier 2008).

Taken together, the qualifications to the "colonial machine" thesis have underlined the essential roles played by the people who lived within and traveled between France's colonies and trading posts and who contributed to development of knowledge about the natural world. Those people, however, often acted relatively independently. Royal coordination and financial aid were never a certainty. Every initiative to seek, collect, and study new specimens was worked out among an assortment of patrons and investors whose efforts might, or might not, receive further support. The state was significantly

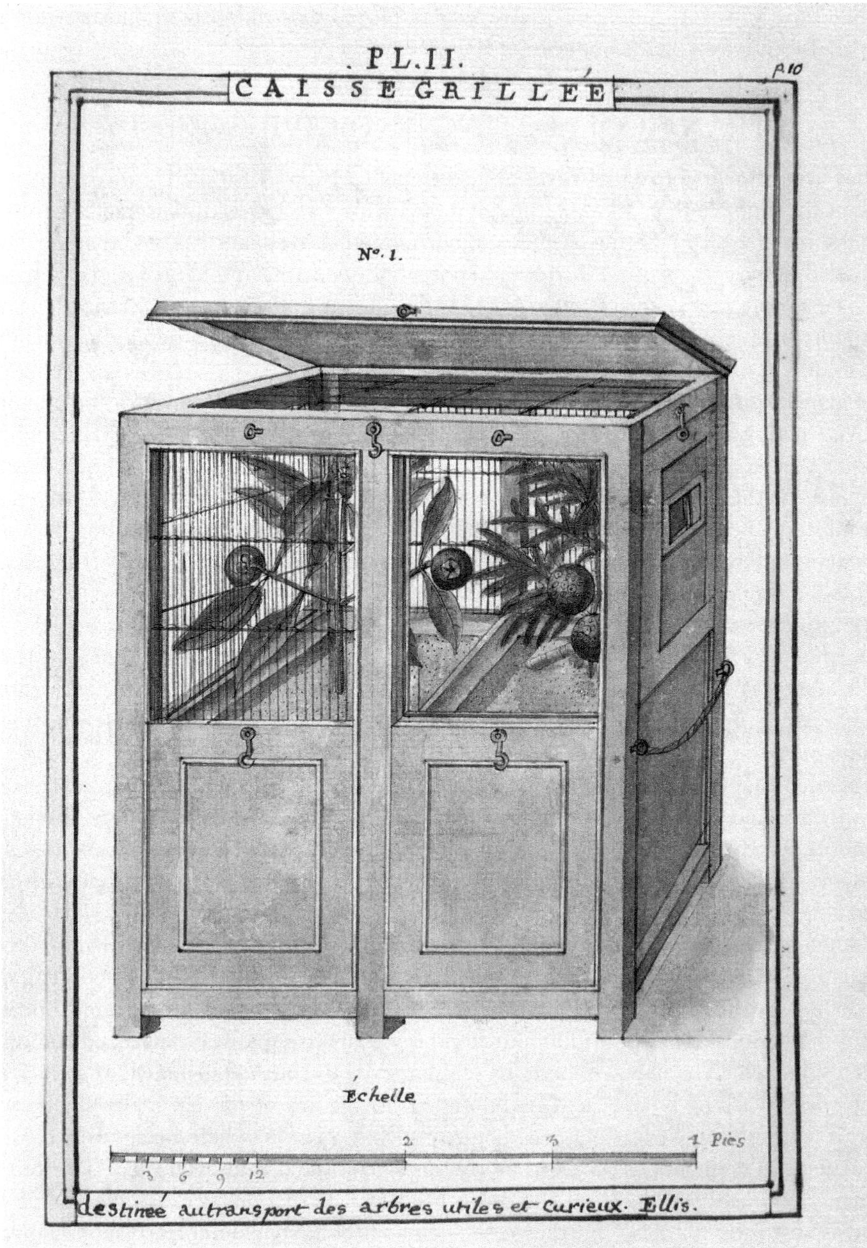

FIGURE 3.2 *Caisse grillée* (Chest for Plant Transport). Watercolor by Delahaye, plate II in *Florindie ou Historiephysico-economique des vegetaux de la Torride* (1789). DEA/G. DAGLI ORTI/De Agostini via Getty Images.

dependent on the goodwill and compliance of its actors. Money was thus not everything: the botanical economy involved multiple, overlapping, forms of value.[6] As the final section here shows, French plant collectors were willing to comply with the directives issued by the Crown in spite of the absence of pecuniary rewards. Such compliance had its limitations, however. Members of civil society played tremendously significant—though largely forgotten—roles in supporting Enlightenment plant collecting.

PLANT COLLECTING AND CIVIL SOCIETY

It has been shown above that royal or governmental support for plant collecting was ad hoc in the eighteenth century. Although it was widely recognized that greater access to natural resources could bring clear economic and imperial advantages, and that increased knowledge about the natural world could potentially generate untold profits, state support was never offered systematically. Instead, the act of collecting and inventorying the natural world fell to a broader public of investigative individuals; a public that included the middling ranks as well as the nobility.

Financial support from the Crown often emerged as a consequence of independent activity taking place in civil society, rather than as initiator of it—thus qualifying the notion of a well-oiled colonial machine. Earlier in the chapter it was mentioned that the botanists at the Jardin du Roi had to petition the Crown to gain support for overseas exploration and botanical collecting in the 1720s. The same structural situation persisted throughout the century. In Nantes in 1765, Francis Bonamy (1710–86) circulated a printed pamphlet addressed to the États de Bretagne, in which he requested an appointment as Professor of Botany (ADLA C632, Bonamy 1765). Bonamy held a chair in medicine at the University of Nantes, and was a member of the Académie Royale des Belles-Lettres of La Rochelle. Bonamy explained that, acting in accordance with earlier orders issued by the King, he had also fulfilled the role of Demonstrator of medicinal plants over the preceding twenty-nine years. This role consisted of offering lessons in the gardens of Nantes and its environs which provided a basic training in medical botany to anyone—but especially those intending to travel overseas. In this respect, Bonamy's example appears to uphold the notion of a colonial machine, especially because, as Bonamy explained, he had dutifully performed this role for several decades without receiving a *sou* in recompense.

Bonamy requested that a Botany School should be founded permanently in Nantes (in emulation of one already created in Rouen), and that he should be appointed Professor of Botany in return for annual *pension* of 1,000 *Louis*. As part of his petition, Bonamy listed the various duties he had undertaken previously to support botanical education in Nantes, for which he now wished to be reimbursed. His role had included, of course, the time spent teaching botany, and he reminded his readers of the particular significance of this form of knowledge to a maritime city such as Nantes, where many of its medical students would embark as surgeons on the ships that sailed from its docks. "The latter," Bonamy emphasized, "have all the more need for instruction, because the life of the crew is entrusted to them." Further costs included the time taken in corresponding with savants, and the expense of obtaining books on botany which, Bonamy emphasized, did not come cheap (ADLA C632, Bonamy 1765).

Although the focus of Bonamy's petition was on obtaining recompense for the time and money he had invested in actually teaching botany, his petition also gestured towards the social economy of botanical research. Through most of the eighteenth century, the collection and study of plants had been supported by wealthy, elite patrons who were considered to be savants. Thanks to their private means and social standing, these individuals could afford to purchase the expensive books that Bonamy mentioned, and had the social connections to be able to borrow other texts from the libraries of their counterparts. They participated in the republic of letters—an international intellectual network (Brockliss 2002: Introduction; Goldgar 1995; Goodman 1994). Characterized thus, eighteenth-century science appears to have been the preserve of the elite and wealthy, a continuation of the status quo from the Early Modern period.

But eighteenth-century botany was at the vanguard of a shift in public participation in science, and this broader participation introduced a different kind of economy for learning about plants. Particularly from the 1760s onwards or, in other words, the decade in which Bonamy wrote his petition and when the large-scale voyages of exploration left European shores, France, Britain, and other European countries saw a significant expansion in public participation in collecting and learning about plants (Cooper 2013; Shteir 1996; Williams 2001). Public interest in botany was facilitated by the fact that the study of the natural world could be very inexpensive: as contemporary introductory books on natural history reminded readers, one simply had to walk into gardens or fields to begin studying the science. There is no doubt, however, that the influx of strange exotic plants from overseas made a significant contribution to stimulating further public interest in botany (Alyon 1787–8; Pluche 1732–42).

A general upturn in consumption in eighteenth-century society further supported plant hunting and the collection of plants, both domestically and from overseas. The expanding consumer society was predominantly an urban phenomenon in France, and the commercial culture surrounding the supply of new, rare plants consequently developed first within cities, especially Paris but also smaller wealthy conurbations such as Nantes. The mounting public interest in collecting and studying rare plants effectively created a new kind of public that supported plant collecting, which comprised primarily the middling ranks and was therefore distinct from that which existed among the upper echelons of society.

The diversification in the social economy of plant collection and study had a twofold effect on the cultural and economic significance accorded to plants. Firstly, public interest was especially orientated towards obtaining ornamental exotics that might make an ostentatious display within gardens. The significance of public interest meant that travelers—who ranged from educated employees of the Compagnie des Indes to military or naval officers, surgeons, and independent merchants—were regularly instructed to bring back to Europe plants that were "useful and beautiful" or "curious" (the term "curious" captured a longstanding association between interesting knowledge and aesthetic appeal) (Duhamel du Monceau 1753; Spary 2004: 14). Throughout the eighteenth century, then, beauty remained a significant criterion, alongside utility, for the plants collected.

The second consequence of the diversified social economy of plant collecting was that the increased demand for new plants stimulated an increase in supply. Plant merchants—known in French as *pépiniéristes* and in English as nurserymen—established commercial gardens in which they gathered together rare, exotic plants for sale. The cultivation of those plants required specialist skills, and the most successful merchants were more than traders: they had to know how to acclimatize plants new to Europe and how to breed specimens that were hardy enough to be sold on. In some cases, the merchants would also selectively breed plants to enhance their aesthetic features. The extent of their ability to do so introduces a fourth kind of value associated with the collection of plants: that of the prestige derived from the ability to cultivate something so delicate and rare. Importantly, the kudos accorded to successful plant cultivation had little to do with the development of taxonomy—the area of enquiry previously assumed to have dominated Enlightenment botanical study (Foucault 1970).

It is worth noting here the significance of this domestic dimension to eighteenth-century plant collecting. Just as the "exploration" central to the collection and study of plants could be either domestic or external, so too was the "trade" on which eighteenth-century plant collecting and transfer depended. The metropolitan nurseries, for example, obtained rare specimens via a combination of commercial trade and gift exchange that otherwise characterized Enlightenment scholarly culture. Although traders might certainly pay a supplier for specimens, they also formed syndicates with networks of scholars and royal institutions, sharing the cost of collecting. Paris' Jardin du Roi thus worked closely with one of the biggest Parisian commercial nurseries, run by married partners Adélaïde

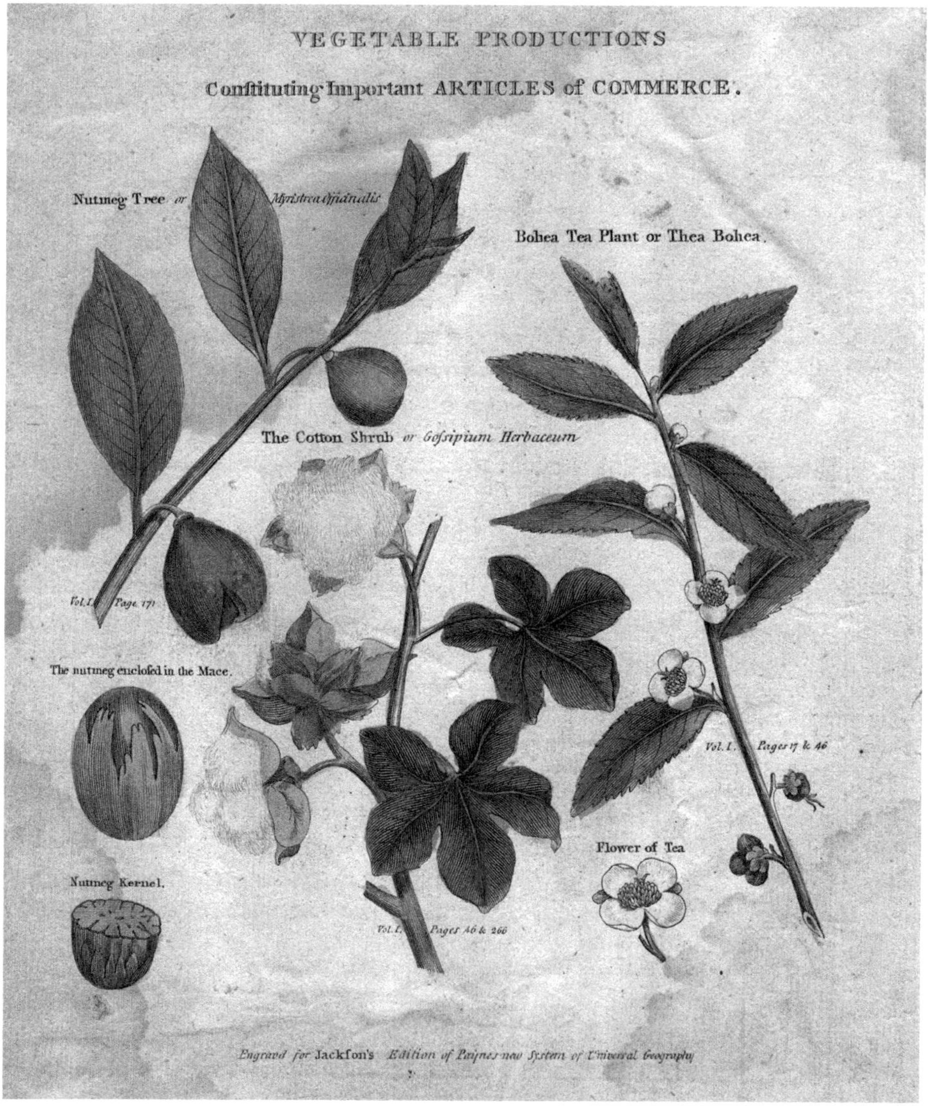

FIGURE 3.3 *Vegetable Productions Constituting Important Articles of Commerce* (*c.*1790). Photo by Hulton Archives/Getty Images.

d'Andrieux (1756–1836) and Philippe Victoire Lévêque de Vilmorin (1746–1804), from whom they bought seeds or live plants to supplement their own collections. The nursery also helped to train the gardeners who were then sent out to collect plants across the world. It was by no means unusual for a plant hunter to work first as an apprentice gardener in a commercial nursery before setting off as a traveling collector (Easterby-Smith 2018: ch. 1). The result of this was that the commercial nurseries had direct access to specimens sent to them by collectors able to anticipate market demand.

The commercial nurseries received exotic specimens sent back to Europe, then, and they did so with the acquiescence of both royal and private patrons. Their gardens thus came to house large collections of rare, exotic plants. While possession of a wide-ranging collection would clearly serve merchants' private commercial interests, it also attracted further interest from scholars of botany. Across the channel in London, for example, Joseph Banks' librarian Jonas Dryander (1748–1810) visited London's Vineyard Nursery 659 times over a twenty-one-year period (Carter 1988: App. XII). Other amateur scholars of botany also recorded going on "botanical tours" of the commercial nurseries around London, where they encountered exotic specimens, such as bananas, for the very first time. The visitors not only viewed the rare specimens on display but also discussed their growth, cultivation, and characteristics with the merchant proprietors. The men and women who worked in these nurseries, in turn, demonstrated a sophisticated understanding of the science of botany, which was linked to their practical ability to cultivate non-native plants. The expanding public demand for exotic plants, then, led to the emergence of a small number of expert merchants who possessed not only the practical skills but also the scholarly expertise to naturalize, cultivate, and study tender, exotic specimens (Easterby-Smith 2018: 21–49, 101).

Eighteenth-century plant-exchange networks were composed of a mixture of commercial and scholarly individuals amongst whom multiple economies emerged in support of plant-hunting. The forms of value attributed to plant collecting could encompass governments' economic and political objectives, but also the social value and prestige invested in plant collection and study within civil society. Trade, exploration, and scientific research thus went hand in hand both in terms of the ultimate aspirations for plant collecting and in terms of the structures through which plants were obtained.

CONCLUSION

By way of conclusion, this chapter returns to Louis XV's 1726 proclamation quoted at the start of the chapter. The request that plants should be collected by Nantes' merchant navy was largely unsuccessful. Why was this the case? It has been seen that the support offered for eighteenth-century plant collecting ranged widely beyond the remit of the state, and that, in spite of its ostensible advocation of plant collecting, the French government did not offer consistent financial or educational support to uphold its own initiatives. In spite of the high profile accorded to botany by the royal government from the early eighteenth century, botanical teaching and research received minimal financial backing from the government. Transporting plants by sea was very difficult, and it seems that the state offered mariners very few incentives to undertake this challenging task. The merchant navy was not paid to convey plants on its ships, for example, and there was apparently no punishment for non-compliance. The Crown looks to have wrongly assumed that the prestige accorded to successful introduction of new plant species would be sufficient.

To return to the metaphor of the colonial machine, the state mechanism thus lacked the necessary internal cogs and wheels.

This leads to the broader point about the forms of value that supported eighteenth-century plant collecting which were, variously, economic, political, social, or symbolic. The expansion of trade and the increasing emphasis placed on scientific exploration over the course of the eighteenth century was encouraged by, and also further encouraged, the development of these four economies within plant collecting in eighteenth-century Europe. They show, furthermore, that eighteenth-century plant collectors and students of botany were not in fact obsessed with taxonomy—in spite of Michel Foucault's later designation of the Enlightenment as the "age of classification."[7] The growth of a consumer society led to an increase in the consumption of ornamental plants, and this in turn stimulated a new domestic market for exotic species. Close attention was paid to the potential economic value that these new plants might bring. For horticultural collectors at least, the attribution of value was also tied up with judgments about a plant's aesthetic appearance.

A significant change in the approach to plant collecting emerged in the 1760s and over the decades that followed. The 1763 Treaty of Paris, which concluded the Seven Years War, resulted in the redistribution of European colonial possessions and a shift in imperial strategy. France, which saw a significant contraction in the physical landmass under its possession, altered its priorities away from the expansion of formal imperial control, towards "informal" empire founded on integration within non-European markets, and on access to and the extraction of, resources. France also invested in voyages of exploration, which placed great emphasis on collecting and studying flora and fauna. By the latter third of the eighteenth century, too, the number of colonial botanical gardens established by European powers had increased to the extent that a network now spanned the globe.

These large-scale activities, initiated either by governments or by trading companies, ran alongside the rising public interest in collecting plants. Again, the 1760s mark a key turning point within a longer-term narrative. The new media attention paid to the initiatives discussed above encouraged greater public interest in plant collecting, which was then made possible in a practical sense because of the expansion of consumer practices among the wider public. The rise in consumption had a substantial effect on trade, exploration, and the collection of new plants. The growing demand from consumers for new horticultural plants in turn encouraged the development of networks of plant traders who exchanged specimens with each other and who trained and directed the plant hunters sent overseas. This resulted in the creation of a greater number of personnel with the necessary skills to facilitate plant transfer. The expanding public interest in plant collecting also encouraged a growth in the provision of public botanical education, which was in itself the product of a broader shift in attitudes towards encouraging curiosity and intellectual enquiry.

From its inception under Louis XIV, French natural history emerged as a discipline directly associated with notions of utility and economic development. But state support for botanical travel, trade, and exploration was neither natural nor automatic in the eighteenth century. Private patrons and the wider public also played significant roles, acting as advisors or sponsors for botanical travel and plant collection. While the impetus for collecting and studying plants often came from the Crown and its institutions, the details of that initiative were just as often worked out within civil society.

CHAPTER FOUR

Plant Technology and Science

ALEXANDRA COOK

> Indigenous peoples offer us ... a skill that amazes us. The benefit they derive from the majority of their countries' plant productions for various purposes appropriate to their way of life ... proves at the same time how much these peoples surpass us in the knowledge of the true properties of plants.
>
> (Toscan 1800: 54)

> ... they became so overwhelmed with the multitude of species, as almost to despair of finding the way in or out of their gardens; both the [East and West] Indies[1] daily furnishing them with so many novelties, that no memory was strong enough to retain them.
>
> (Linnaeus [1761] 1786: 3)

INTRODUCTION

Writing just before the end of our era, Georges Toscan (1756–1826) of the Muséum national d'histoire naturelle (National Museum of Natural History) in Paris, summarized three centuries of European prospecting by praising indigenous peoples for *surpassing* Europeans in their "knowledge of the true properties of plants." Toscan was ahead of his time in recognizing the often superior quality of indigenous knowledge, thereby giving credit where it was due.

Whether it was the pineapple, the potato, tomato, corn, tobacco, quinine, or chocolate from the New World, coffee, okra, or rice transferred from Africa, or tea, ginseng, and the "true" rhubarb from China or New Zealand flax, Europeans everywhere and always depended on local people to name and explain how to use these plants (Eamon 2018; Schiebinger 2004b). While usually neglected in traditional accounts of the history of plant science and technology, indigenous and subaltern knowledge, including classification, shaped modern plant knowledge in crucial ways, with many modern plant names traceable

The author gratefully acknowledges generous funding from the Research Grants Council of the Hong Kong Special Administrative Region, China (General Research Fund Grant 17401614 and Humanities and Social Sciences Prestigious Fellowship and Project No. HKU 17401614), the Institute for Advanced Study, Nantes, France, and research assistance by Johannes E. Hoerning, Nikolaj Bichel, and Felix Yeung Shing Hay. All translations are the author's unless otherwise indicated.

FIGURE 4.1 Pineapple depicted by Maria Sibylla Merian in *Metamorphosis Insectorum Surinamensium/Ofte Verandering der surinaamsche Insecten* (Amsterdam: Gerard Valck, 1705), pl. 2. Royal Collection Trust/© Her Majesty Queen Elizabeth II 2020.

to non-Western languages.[2] This contribution was due not only to indigenes' profound knowledge of plants, but also to the sheer number, usefulness, and diversity of plants in regions known today as "biodiversity hotspots."

Biodiversity hotspots cover less than 3 percent of the earth's surface, yet contain approximately 50 percent of its plant species. While providing highly coveted spices and other valuable commodities to Europeans, this cornucopia proved overwhelming, spawning a peculiarly modern problem: an information explosion (Slaughter 1982: 10). It was as if one wanted—in the absence of a catalog—to find a particular book in a library of thousands of items, with each book having multiple titles. Finding a way out of this predicament demanded greatly enhanced information management and retrieval methods—efficient and unambiguous plant nomenclature, coherent classification systems, and functional formats for specimen storage. Without these three *sine qua nons*, all other work on plants—whether physiological, experimental, pharmaceutical, or horticultural—would have been greatly hampered or even impossible. Hence, these three pillars of seventeenth- and eighteenth-century plant science and technology form the major focus of the present chapter.

By emphasizing information management and retrieval, as well as their epistemic consequences, this chapter foregrounds the immense and uniquely European effort to create an unprecedented knowledge artifact: a truly *global* catalog of all plants known in Europe, including many from coastal areas of Africa, Asia, and the Americas. Before Caspar Bauhin (1560–1624) listed 6,000 plants with their often multiple synonyms (Dietz 2019) in his *Pinax theatri botanici* (1623) no effort of a comparable scale had ever been undertaken anywhere. While plant data management challenges remained small by twenty-first-century standards, their novelty and scale were at the time daunting. In order to gain a sense of the magnitudes involved, consider a typical "folk" taxonomy, which, *irrespective of culture or geography*, comprises roughly 500 to 600 species-type units, and no more than 1,000 basic plant kinds (Berlin 1999: 78; Stevens 2002: 19). Renaissance herbals conform to this pattern, as does Li Shizhen's compendium of *materia medica*, *Bencao Gangmu* 本草綱目 (1596) (Atran 1990: 18).[3] The numbers rose dramatically from this point. Rembertus Dodoens' *Cruydeboeck* (1554), discussed 1,340 species, of which 600 were reported for the first time,[4] while Francisco Hernández de Toledo (1515–87) collected information on 3,000 New World plants (*c.* 1576). After Bauhin published 6,000 species in *Pinax*, John Ray (1627–1705) reported 18,000 species in *Methodus plantarum* (1682), even though divergent taxonomic criteria may account for some of the difference (Cailleux 1953: 45–6; Crombie 1994: 1263). The increase in the number of plants known to Europeans had therefore long since burst the confines of traditional classifications by the time Carolus Linnaeus (1707–78) wrote of the "despair" occasioned by so many plants arriving from the "Indies" (Linnaeus [1761] 1786: 3).

The ensuing information management breakthroughs introduced by Bauhin, Linnaeus, and others laid the groundwork for a seismic shift in the epistemological status of plants as a domain of study in their own right, independent of medicine, pharmacy, and chemistry. Like other technologies that developed rapidly during this era, such as shipbuilding, navigation, and ballistics, plant data management ultimately promoted major reorganizations and transformations of knowledge (Black 2014).

However, while throwing into relief the unprecedented information management challenge confronting Early Modern Europeans as they plundered, regulated, and traded the products of "biodiversity hotspots" across the globe, the present chapter likewise follows Toscan in decolonizing the plant knowledge so often claimed by Europeans as their own. Throughout what follows it is therefore crucial to bear in mind that cataloging and using the plant riches of the "Indies," whether western or eastern, always depended on local people; at the same time this project was intimately bound up with maintaining control over non-Western peoples and resources, and would remain so for centuries to come.

SORTING THROUGH THE BABEL OF BOTANICAL NAMES

In order to think about, use, and interact with things in the world, ways are needed to distinguish them from one another. Naming is therefore a foundational epistemological activity. "Folk," that is to say, ordinary people, practice first-order nomenclature worldwide: they give different names to things they take to be different, while giving the same name to things they consider to be the same. The Bible offers a well-known model for such first-order naming, which was invoked by Linnaeus: Adam was charged by his Creator with naming the plants and animals in the Garden of Eden (Gen. 2:19–20). Folk typically employ a basic set of about five hundred names to create two-word combinations numbering in the thousands that account comprehensively for the plants of a given region (Berlin 1999: 78).

Folk plant nomenclature derives from varied sources; names may refer to anthropocentric uses (medicinal, alimentary, and ornamental), plant characteristics, e.g. gross morphology, size, aroma, taste, smell, habit, sex, or geographical origin, among others (Atran 1999: 128).[5] This plurality characterizes nomenclatures as disparate as Tzeltal Maya (Mexico) (Atran 1999: 128, 155) and Chinese. Yet, while plural, these categories are often similar across cultures. For example, in *Bencao gangmu* there is "an exact parallelism between East and West in the choice and construction of plant names ... every one of the categories into which the Chinese phytonyms were divided has its counterpart among the Western ones" (Needham et al. 1986: 165; see also Elman 2007: 146; Métailié and Hsu 2001).[6] However, contrary to what is often asserted, folk names do not necessarily reference medicinal uses since medicine in many cultures constitutes closely guarded secret knowledge (Cook 2010: 129; Kathirathamby-Wells 2015; Mahr 1955: 11–12; cf. Schiebinger 2004b: 196, 220–2).

When Europeans ventured onto the high seas in the fifteenth century in search of new trade routes, spices, drugs, and precious metals, they encountered a huge variety of plants, with their local names, ranging from Taíno and Arawak in the Caribbean to Dravidian languages of South Asia, to Sanskrit, Chinese, Japanese, Malay, and other languages of the East Indies. Although these novelties might have been less accessible in cultures of secrecy, Europeans nonetheless became aware of indigenous plant names. As Toscan observes, "Australian natives have given specific names to each plant; and ... their knowledge in botany is invariable" (Toscan 1800: 54).

Nearer to home, the German physician, Leonhard Rauwolf (1535–96), made a three-year voyage to the Near East in search of medicinal plants for European consumption.[7] His journey highlighted urgent problems of synonymy and referential ambiguity based in the erroneous belief of "many scholastic 'naturalists' ... that the local flora of the temperate regions of Europe could be exhaustively apportioned among the Mediterranean plants depicted in ancient sources" (Atran 1990: 127). On his epic journey Rauwolf described 364 plants and collected 200 specimens, many labeled with their names in Arabic and/

or local dialects (Ghorbani 2018). Like all such collectors, Rauwolf had then to align indigenous names and specimens with synonyms in order to avoid costly and risky errors.

Faulty translations from and among canonical works could likewise generate mistakes in plant names:

> … rendering of plant names from Greek and Arabic into Latin was accomplished in some instances by translation, in some by transliteration; different authors might choose different ways of rendering the same name. … the same plant might be known by several different names, or a completely different plant might be identified with one mentioned by an ancient or Islamic author.
>
> (Siraisi 1990: 143)

The synonym problem—confusion caused by plants having multiple names—dominated seventeenth- and eighteenth-century botanical work (Dietz 2019). Concern with this problem was driven by medical and commercial motives: in the era before synthetic drugs it was imperative for botanists, who were predominantly physicians, surgeons, or apothecaries, to be able to identify medicinal plants. Using the wrong plant could be ineffective or worse: fatal. China root (*Smilax china*) and quinine or quina (*Cinchona officinalis*) are well-known examples of unstable pharmaceutical names applied to different plants at various times and places (Boumediene 2016: 193; Winterbottom 2014). Druggists (not to be confused with apothecaries), trading companies, and merchants likewise depended on certain knowledge of plant identities: were they, for example, trading in the true cinnamon or a less-prized species of the same genus (Garcin et al. 1742)? Tea, which appeared in green and black forms, might derive from one species or, according to Linnaeus, from two. Given tea's high market value, much was at stake in obtaining the correct answer to this conundrum. The same applied to all manner of plant-derived commodities such as spices, woods, and so on.

Researchers matched descriptions, images, and specimens to features such as flowers, leaves, reproductive organs, and cotyledons (embryonic seed leaves) in a visual and quantitative way in order to link each plant with one name and its corresponding synonyms. In the case of tea, Linnaeus' two-species theory was refuted by counting the petals of "several hundred flowers" and finding their number to be consistently the same (Lettsom 1772: 7). Plant physiology, then an emerging field of study, played therefore virtually no role in establishing this catalog of plant identities. Hence, a contemporary textbook concluded that "Botanists' research should essentially concern only [plants'] *exterior parts*. The examination of internal organs belongs to the *Physicist* who seeks to discover the laws of vegetation" (Latourrette and Rozier 1766: 24, emphasis original).

The synonym problem, and its corollary, the homonym problem—one name shared by different plants—has by no means vanished over time; many pharmaceutical publications mistakenly use outdated or invalid synonyms (Dauncey et al. 2017). Plant identity is crucial in Ayurveda, naturopathy, and Traditional Chinese Medicine (TCM), where actual plant materials (versus synthesized bioactive compounds) are used. Modern confusions may concern vernacular names such as "ginseng," which designates twelve plant species from six different genera, each with its own chemistry and therapeutic properties. In one serious case, a hundred people needed life-long dialysis due to kidney damage caused by a toxic plant that shares the name of a TCM plant (Wills 2017: 281; see Lei 2014: 208 concerning TCM homonym, *chang shan* 常山). Stable nomenclature therefore remains a prerequisite for all work with plants, whether taxonomic, experimental, horticultural, or pharmaceutical.

Another difficulty was posed by the sheer length of European plant names, often comprising long descriptions that have been considered cumbersome and of low scientific utility. Yet such names afforded a degree of inclusivity now lost to science, incorporating synonymies, traditional names, and descriptive information. *Aureliana canadensis sinensibus Gin seng Iroquaeis Garent oguen*, Father Joseph-François Lafitau's (1681–1746) polynomial for ginseng, exemplifies this inclusivity. The name he created includes both ginseng's Chinese name, ginseng (*ren shen*), referring to "man-root," and its Iroquois name, *garent oguen*, referring to "human thighs" (Cook 2010: 129). Similarly, this name refers to the origins of different ginseng species in Canada and China.[8]

Compare Lafitau's name with the modern name for the Canadian ginseng species, Linnaeus' *Panax quinquefolius*. The latter name shows no connection either with the Iroquois, their name for ginseng, or Canada. As will be shown, nomenclatural inclusivity is not a trivial matter, given the pared-down binomial structure of modern scientific names and the many ways in which indigenous sources of plant knowledge were not necessarily erased, but quite often given less than due acknowledgment (Raj 2017; cf. Schiebinger 2004b).

Bauhin, a leading compiler of plant data, made important contributions to nomenclature by creating a "name [that] is the essence of the thing and constitutes a polynomial of variable structure" (Selosse 2008: 77). Yet his polynomials exemplified the practical limits of scholastic naming practices; by the middle of the eighteenth century, new record-keeping practices introduced by Linnaeus produced a paradigm shift in European plant nomenclature that, while having proven its staying power, nonetheless sparked controversies which haunt scientific nomenclature down to the present.

Responding both to the natural history "information explosion" and unwieldy polynomial plant names, Linnaeus portrayed himself as a second Adam, conferring plant names as if conducting a religious reform (Linnaeus 2003: 15; Müller-Wille 2007: 545).

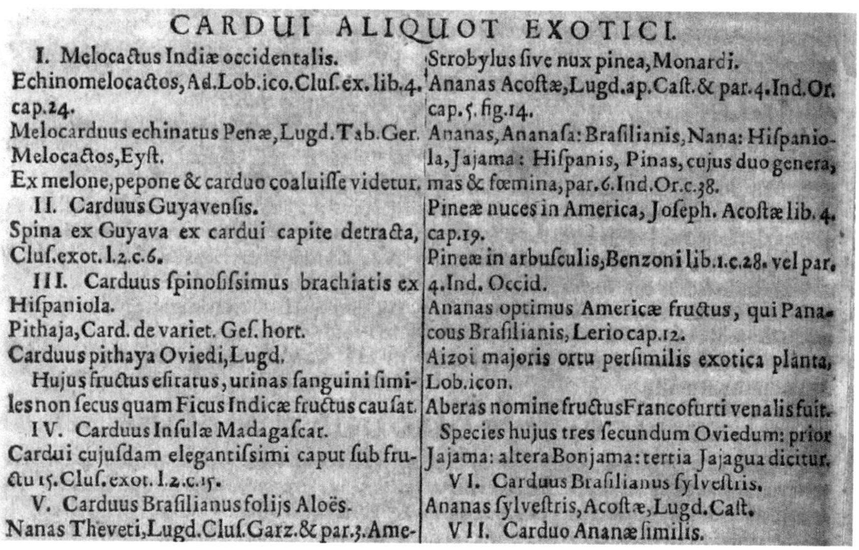
CARDUI ALIQUOT EXOTICI.

I. Melocactus Indiæ occidentalis.
Echinomelocactos, Ad.Lob.ico.Cluſ.ex. lib.4. cap.24.
Melocarduus echinatus Penæ, Lugd.Tab.Ger.
Melocactos, Eyſt.
Ex melone, pepone & carduo coaluiſſe videtur.
II. Carduus Guyavenſis.
Spina ex Guyava ex cardui capite detracta, Cluſ.exot.l.2.c.6.
III. Carduus ſpinoſiſsimus brachiatis ex Hiſpaniola.
Pithaja, Card. de variet. Geſ. hort.
Carduus pithaya Oviedi, Lugd.
Hujus fructus eſitatus, urinas ſanguini ſimiles non ſecus quam Ficus Indicæ fructus cauſat.
IV. Carduus Inſulæ Madagaſcar.
Cardui cujuſdam elegantiſsimi caput ſub fructu 15. Cluſ. exot. l.2.c.15.
V. Carduus Braſilianus folijs Aloës.
Nanas Theveti, Lugd. Cluſ. Garz. & par.3. Ame-
Strobylus ſive nux pinea, Monardi.
Ananas Acoſtæ, Lugd.ap.Caſt.& par.4.Ind.Or. cap.5. fig.14.
Ananas, Ananaſa: Braſilianis, Nana: Hiſpaniola, Jajama: Hiſpanis, Pinas, cujus duo genera, mas & fœmina, par.6.Ind.Or.c.38.
Pineæ nuces in America, Joſeph. Acoſtæ lib. 4. cap.19.
Pineæ in arbuſculis, Benzoni lib.1.c.28. vel par. 4.Ind. Occid.
Ananas optimus Americæ fructus, qui Panacous Braſilianis, Lerio cap.12.
Aizoi majoris ortu perſimilis exotica planta, Lob.icon.
Aberas nomine fructus Francofurti venalis fuit.
Species hujus tres ſecundum Oviedum: prior Jajama: altera Bonjama: tertia Jajagua dicitur.
VI. Carduus Braſilianus ſylveſtris.
Ananas ſylveſtris, Acoſtæ, Lugd. Caſt.
VII. Carduo Ananæ ſimilis.

FIGURE 4.2 Entry on pineapple, "Ananas," and its synonyms. Caspar Bauhin, *Pinax Theatri Botanici* (1623: 384). Courtesy of Biodiversity Heritage Library.

In his works he prescribed rules for naming genera, the supra-species category that he considered essentially natural (Linnaeus 2003: 163, §229). The best generic names, he opined, "show the essential character or habit of the plant" (191); yet Linnaeus has become notorious for claiming that generic names memorializing botanical heroes "should be religiously preserved" (185, §238). Theoretically, the seeming conflict between these two rules is resolved by the stipulation that "[s]uch names are to be bestowed when *no distinguishing character* readily suggests itself, and there is no suitable alternative" (Linnaeus 1938: 71, emphasis added).

Another important Linnaean rule for plant nomenclature is that "[g]eneric names, which do not have Greek or Latin roots" must be excluded, and "[a]ll barbarous names are regarded by us as primitive, since they are from languages not understood by the learned. [So are] doubtful appellations of plants, when it is hard to decide what language they are derived from" (Linnaeus 2003: 172, §220). As will be shown, these Eurocentric naming requirements have garnered much attention—often negative—both in the past and in the present. On the other hand, Linnaeus felt by no means bound to adhere to his own rules, violating them as he saw fit (Cook 2010).

In scientific naming Linnaeus is well known for a simple, yet effective innovation that he began using in the 1740s: the species label or "trivial name" (*nomen triviale*) appended in the margin next to the generic name and polynomial species description. This ingenious shorthand originated "as an indexer's paper-saving device" (Åsberg and Stearn 1973: 6) or "paper tool" (Müller-Wille 2017: 116) during Linnaeus' work on *Öländska och Gothländska Resa* (1745) and *Pan Suecicus* (1749).[9]

The trivial name is purely *arbitrary*, serving as a mere label, signifier, or "baptismal name" conferred in a "dubbing ceremony" (Kripke 1971; Waxman 1999). Unlike the generic name, this baptismal name does not convey anything *essential* about the species being named; rather, it is "*a word freely taken from any source*" and "free from any laws" (Linnaeus 2003: 219, emphasis original). For Linnaeus the trivial name therefore had a fundamentally different function from the true plant name, i.e. generic name joined to species description.

Linnaeus' binomial nomenclature developed incrementally over several years, with its application in *Philosophia botanica* (1751) still "scattershot and at the mercy of hurry and chance" (Koerner 1999: 53). Linnaeus first used binomials consistently in *Species plantarum* (1753), and in 1905 the international botanical community accordingly enshrined 1753 as the starting date of modern botanical nomenclature for most groups of plants.

Given the plethora of synonyms and polynomials, binomial names were intended to ensure that species so designated would *always and everywhere be the same* (Kripke 1971: 213). Thus, "[b]y the time Linnaeus died in 1778, most botanists were using binomials as *the* species name" (Stevens 2002: 13, emphasis original). Even those such as Antoine-Laurent de Jussieu (1748–1836), who followed different principles of classification than those advocated by Linnaeus, adopted binomials without hesitation (Jussieu 1778: 178, 193; 1808: 11–12). Modern plant names feature a third, required element, the internationally agreed abbreviation of the name of the botanist to whom the species name is attributed, e.g. "L." for Linnaeus (*Lilium candidum* L.). In its current form, the official plant name therefore consists of (a) generic name, (b) species name (formerly trivial name), and (c) authority. The third element of this technical convention is not considered necessary in general works such as *The Cultural History of Plants* and is therefore not used in this book (see the Series Preface).

Linnaean binomials revolutionized botanical data management, trumping viable alternatives from other European botanists of the Enlightenment, such as Gaspard Bauhin (1560–1624), Joseph Pitton de Tournefort (1656–1708), Albrecht von Haller (1708–77), and Michel Adanson (1727–1806). Binomials made botany, already becoming a popular science, more accessible, and easier to learn because the binomial structure is rooted in human cognition: "The noun + adjective combination in Linnaeus' *Species plantarum* is similar to comparable combinations in folk taxonomies" (Stevens 2002: 19). For Jean-Jacques Rousseau (1712–78) Linnaeus' binomials are "as convenient and necessary for Botanists as ... Algebra is to Geometers" (Rousseau 2000: 97) and "are the only ones admitted in all of Europe and by means of which one is certain to be able to arrive at mutual understanding with botanists of all nations" (172). Jussieu, the doyen of late-eighteenth-century Paris botanists, concurred: "For the nomenclature of Tournefort, the only one admitted until then in the [King's] garden, [I] substituted that of Linnaeus, more abbreviated, more convenient, and generally adopted in all of Europe: which harmonizes the [King's] Garden of Paris with other foreign botanical gardens" (Jussieu 1808: 11–12).[10] Binomials were therefore widely adopted within twenty years of the 1753 publication of *Species plantarum*.

Binomial nomenclature also prevailed due to its adaptability: "The binominal implies neither a belief in a Linnaean ontology, essentialism, or some outmoded paradigm ... it is *theory independent*" (Stevens 2002: 19, emphasis added); since binomials do not depend upon a specific classification system, they may—as Antoine-Laurent Jussieu demonstrated—be used in conjunction with *any* classification system.

Nonetheless, advocates expressed reservations about Linnaean nomenclature. Rousseau, considered a major popularizer of Linnaeus, ultimately rejected Linnaeus' conception of the botanist as the second Adam, the master name-giver: "I have always believed one could be a very great botanist without knowing a single plant by its name" (Rousseau 2000: 131). Jussieu echoed this attitude: "For a long time the definition of botanical science has been incorrect, deceptive, or at the very least, vague; it has concentrated on designating and naming plants ... as a science of pure nomenclature ... " (1789: xxxiv–xxxv; Stevens 1994: 354).

LATIN NOMENCLATURE ON TRIAL

> Let all the other languages *of Europe* be banished from this science, as well as all the languages spoken *outside Europe*, which to us are "barbarous" ...
>
> (Linnaeus 1938: 38, emphasis added)

Linnaeus' privileging of Latin nomenclature was controversial in his own time and remains so to this day. Even though Latin had been the dominant scientific language since antiquity, the most important works on plants from outside Europe relayed a wealth of indigenous names (Hernández and Recchi 1651; Reede 1678–93; Rumphius 1741–50). Linnaeus himself relied on Reede's *Hortus Malabaricus* as an essential work. From the outset, opponents of Linnaeus' language policy advocated traditional names, both European and others: these opponents included Michel Adanson (1727–1806), one of the few European savants of this period to study African flora (1749–53) (Adanson [1763–4] 1966: clxxii–iv); Albrecht von Haller (1708–77), a towering polymath and author of the first comprehensive Swiss flora (Haller 1768); Johann Amman (1707–41), a Swiss botanist employed in St. Petersburg; and the Creole priest, José Antonio Alzate y Ramirez (1737–99) (Drouin 2005: 54). Others such as Jean-Baptiste Christophe Fusée Aublet (1720–78) included Creole and slave plant names alongside Latin binomials and French names.

More recent critics have indicted Linnaeus for "linguistic imperialism" (Needham et al. 1986: 19, 168; Schiebinger 2004b: 194–225) and "symbolic violence" (Drouin 2005: 53; Drouin 2015) because Linnaeus considered it "a religious duty" to memorialize great (usually white, male, European) botanists in generic names. According to these authors, Linnaeus thereby eliminated indigenous names as well as indigenous knowledge. Yet Linnaeus stated that optimally generic names should derive from "the plant's essential character, or its appearance" (Linnaeus 1938: 79). In theory, then, eponymous names presented a last resort.

The "linguistic imperialism" thesis has shifted attention away from Linnaeus' important contributions in preserving indigenous plant names and uses; similarly, his insistence that only *botanists* require Latin names has been ignored (Linnaeus 1938: 37–8). Linnaeus' project—more open and flexible than his early works might suggest—operated at many levels, with indigenous names and knowledge playing an indispensable role: "In his Latin works [Linnaeus] listed vernacular plant names. He made his students do the same, arguing that local names reflect unknown uses or properties of plants" (Koerner 1999: 100).

Many Linnaean generic names in fact reflect *cross-cultural borrowing* and/or refer to traditional medicinal, artisanal, or alimentary uses. Manilal points to *Species plantarum*'s key role in preserving 258 names derived—via Reede's *Hortus Malabaricus*—from Malayalam, a language of southwest India (Tamil Nadu) (Manilal et al. 2003). An analysis of 286 of the 1,313 Linnaean generic names in *Species plantarum* reveals that 11 percent derive from "barbarous" languages including *Coffea* (Arabic), *Spinacia* (Persian), *Saccharum* (Sanskrit), and *Ananas* (Tupi), to name just a few (Cook 2010: 128).[11] Linnaeus subsequently published the Chinese- and Japanese-origin name, *Ginkgo* (*G. biloba*) (Linnaeus 1771: 313).

Beyond the direct adoption of local plant names, Linnaeus preserved indigenous knowledge by stipulating Latin names that record local *uses* and concepts. This important contribution to preservation of indigenous plant knowledge has often been completely overlooked in the literature. Such names include *Panax*, referring to ginseng's widespread reputation as a panacea, and *Sapindus*, referring to plants traditionally used for soap in Florida and the Caribbean (Cook 2010: 129–31).

Other criticisms target Linnaeus' reliance on botanical Latin, a written language derived from a classical language that has nonetheless "in consequence of its neutrality, become world-wide" (Stearn 1959: 9). Latin's role in botany raises two issues: first, Latin's European origin and second, its role as a language of educated elites that effectively restricts participation in science. With reference to the first, Needham et al. (1986) and Schiebinger (2004b) consider modern scientific nomenclature to be fatally tainted by Latin's implicit connections with slavery, colonialism, and disdain toward non-European Others.

With respect to the charge of elitism, Latin's widespread use in science, as in many other disciplines, undoubtedly contributed to gender discrimination, since even literate women usually did not learn it (Schiebinger 2004a: 200; Stearn 1992: 7). On the other hand, many female plant specialists acquired proficiency in botanical nomenclature and botanical Latin more broadly (Benharrech 2018: 91; Meeker and Szabari 2016; Rousseau 2000: 172; Shteir 1995; Weisberg-Roberts 2009). These included Madeleine Françoise Basseporte (1701–80), Madeleine Catherine Delessert (1747–1816), Mary Delany (1700–88), Margaret Cavendish Harley Bentinck, Duchess of Portland (1715–

85), Queen Charlotte of Great Britain (1744–1818), Priscilla Wakefield (1751–1832), Empress Josephine of France (1763–1814), and Maria Edgeworth (1767–1849).

CLASSIFICATION: IN THE BEGINNING THERE WAS THE FOLK

Michel Foucault claimed that the desire for a universal science of order inspired the Early Modern European "classical age" (Foucault 1970). Yet classification is an ancient, "culturally universal" practice: *all* known cultures classify as well as name the living beings in their immediate environment: plants, animals, birds, reptiles, fish, and generally do so comprehensively (Berlin 1999).

The classifications developed by a group of people within particular ecological and cultural contexts are known as folk taxonomies. These typically display nested categories of about three ranks, which correspond to what people can remember: "The general structure of the Linnaean hierarchy is that of folk taxonomies worldwide ... they include 600 or fewer species-type units ... The size and structure of such hierarchies ... is such that they are easy to memorize" (Stevens 2002: 19).

It is usually assumed that folk are most concerned with "salient" (i.e. useful) species (Atran 1999). For example, New World plants introduced in the Philippines were "adopted and named by the indigenes, who in many cases, seem to have even rediscovered their medicinal usages, rigorously parallel to the traditional Mexican ones" (Lévi-Strauss 1962: 22–3). Utility alone, however, does not explain why people create fine-grained classifications of their plant world. Lévi-Strauss argued that folk's "first concern is not of the practical sort. It responds to intellectual exigencies before, or instead of, satisfying needs" (16). In indigenous cultures of Mexico and Peru, for example, "more than a third of the named plants ... have no known social use, nor are they poisonous or pestiferous" (Atran 1990: 20). Categorization is often based on shape, color, or other features having little or nothing to do with use: in "northern Luzon (Philippines) ... major sub-categories of rice are based on overall plant morphology and then on the glutinous or waxy quality of the endosperm" (Conklin 1980: 20).[12]

A major issue in the recent research into folk taxonomies is whether these exhibit affinities with modern scientific taxonomy: one noted anthropologist holds "that when scientists and folk carve up the biological world, they tend to make the same basic cuts" (Atran 1999: 156). Convergence between folk and scientific classifications often results from morphological and other group similarities (Berlin 1999: 71). For example, on the one hand, slaves transported from Africa to Surinam in the eighteenth century employed morphological reasoning to establish a new supply of essential medicinal and comestible plants based on their knowledge of African flora (van Andel 2015). On the other hand, an important philosopher of biology sees folk classifications as being on a par with scientific ones, but making *different* cuts than scientists do (Dupré 1999: 461).

EUROPEANS' SEARCH FOR ORDER

> The "Ariadne's thread" of botany is system without which botany is chaos.
>
> (Linnaeus 2003: 113)

Size is a typical category in folk taxonomies (Atran 1999: 156) and forms the basis for Europe's most long-lived classification: Theophrastus' (372–*c.* 287 BCE) division of plants into "tree, shrub, under-shrub, [or] herb" (Theophrastus 1916: 23). The ancient Greek philosopher's categories remained in use for nearly two thousand years, applied as late as the seventeenth century in landmark works by Bauhin and Tournefort (Greene 1983: 948) and to the twentieth-century system of Kew botanist, John Hutchinson (1884–1972). Apart from size-related categories, Renaissance herbal writers[13] listed plants alphabetically and derived ordering principles from medicine, sometimes arranging plants by the diseases they could treat. Such distinctions had limited value, however, in the face of nomenclatural confusion, lack of consistent classificatory criteria, and above all, overwhelming amounts of information.

Bauhin's *Pinax* presented an unprecedented register of plants, both native and imported, then known to Europeans, comprising six thousand species, descriptions, and synonymies to other authors. The work's utility for data management was furthered by "paper tools" such as indices, lists of authorities, and abbreviations. Consequently, *Pinax* served as a common point of reference for Tournefort, Linnaeus, and those who used their works. Even after Linnaean binomials supplied rigid designators, disagreements about synonymy were, as Rousseau noted, still referred to Bauhin (Dietz 2019; Rousseau 2000: 233).

It is commonly thought that *Pinax* was nothing more than a mere register (its literal meaning), and that its allegedly "incomprehensible order" (Magnin-Gonze 2009: 86) did not advance plant classification (Hoquet 2005: 156). Yet *Pinax* grouped related species together and adopted the kinds of empirical techniques and collaborative work that distinguished Renaissance and Early Modern plant study and science more generally from their predecessors. Bauhin accordingly used not only traditional written sources, but also dried plant specimens, many sent to him by his large correspondence network, a key element of the doing of Early Modern plant study (Benkert 2016). In addition to these epistemic features, *Pinax* broke new ground by discarding medical and pharmaceutical concerns "in favour of the concepts of the specialized botanical community" (Selosse 2005: 169). These virtues notwithstanding, *Pinax* could not overcome Early Modern plant information overload; furthermore, non-European plants were relegated to a "chaotic survey" at the end of the work (Heniger 1986: 141).

WHAT INDIGENOUS CLASSIFICATION?

As discussed above, modern scientific plant nomenclature has diverse sources, including indigenous names and usages. Yet indigenous, folk classifications from outside Europe fared less well; only rarely did Europeans retain the taxonomic contexts from the areas they explored, conquered, and exploited. This was not necessarily due to the reluctance of Europeans on the ground to convey traditional taxonomies, but rather to compilers and editors, who usually relied on taxonomies derived from Theophrastus, Renaissance herbals, and Bauhin.[14]

Some Europeans nevertheless penetrated to the taxonomic level in their investigation of novel flora. In the New World, the Spanish quickly recognized the value of local medical knowledge, which led them to study the medicinal and comestible properties of local plants, aided in no large part by indigenous physicians and other knowledgeable local people. In his manuscript, *Historia natural de las plantas de Nueva España* (*c.* 1576), the Spanish physician Francisco Hernández de Toledo (1515–87) presented his massive "protoethnographic" study of three thousand New World plants such as pineapple, tomato, cacao, chili, cacti, and tobacco; aided by local assistants, illustrators, and physicians. Hernández relayed not only illustrations of the plants with their Nahuatl

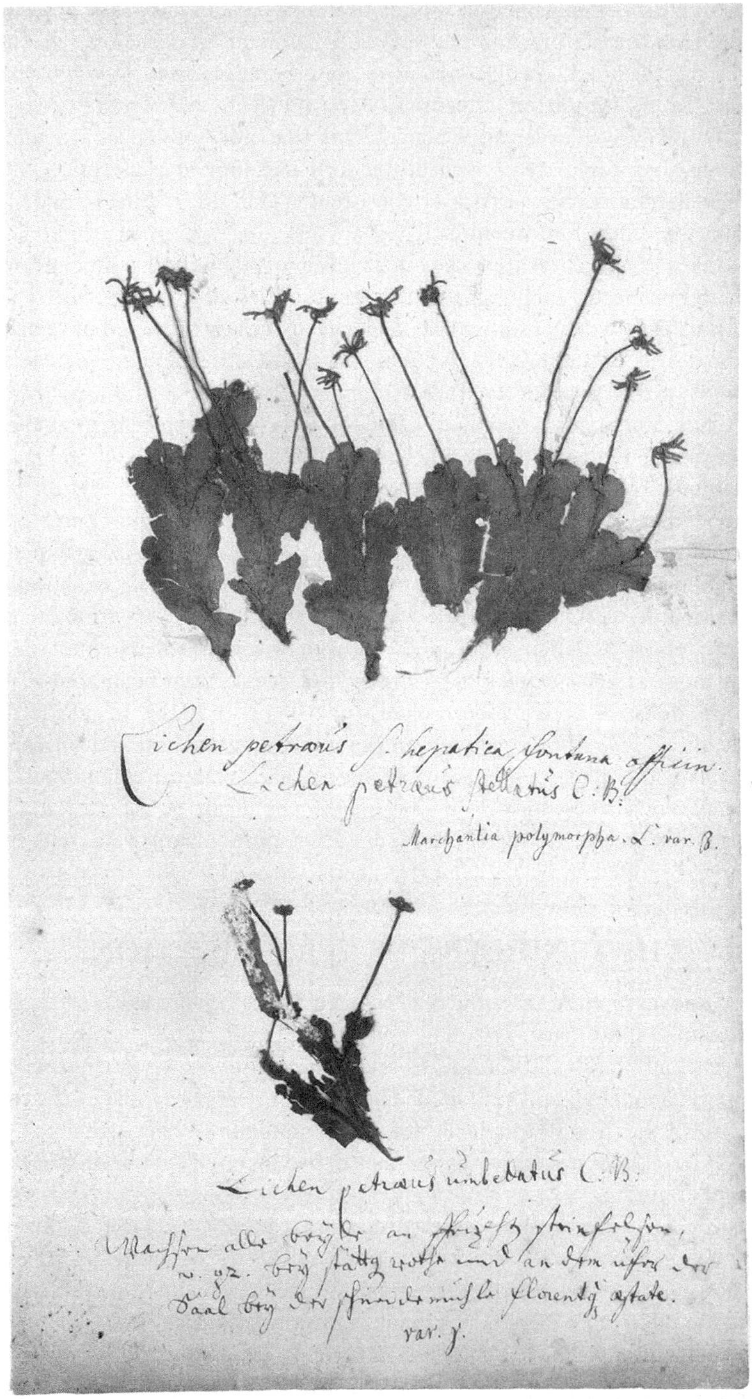

FIGURE 4.3 Specimen of "Marchantia polymorpha." Attributed to the Herbarium of Caspar Bauhin. Reproduced with permission from the Royal Botanic Gardens Victoria.

names, but also indigenous classifications of these plants (Boumediene 2016: 193). Unfortunately, Hernández was unable to see his original work, with its four thousand illustrations of plants, animals, and Mesoamerican gods to publication; instead it passed through several hands prior to posthumous publication in heavily redacted editions such as *Rerum medicarum Novae Hispaniae Thesaurus* (1628), the "most influential of all the works published by the Academia dei Lincei" in Rome (Arredondo 2019: 46; see also Hernández and Recchi 1651). None of these published versions included the indigenous classifications cited by Hernández.[15]

By the time Nicolaus Joseph (later Freiherr von) Jacquin (1727–1817) explored Caribbean flora in the mid-eighteenth century, Linnaean names and taxonomy had become almost *de rigueur* (Jacquin 1763). While his multilingual flora of French Guiana included Creole and slave plant names and uses, Aublet's classification conformed to the Linnaean sexual system (Aublet 1775). Any family resemblances or other folk classifications at work therefore failed to be recorded even in one of the most culturally and linguistically inclusive works of the period.

NARDI ANT. RECCHI LIB. VIII. 312

De MATZATLI, ſeu Pinea Indica. Cap. LXXIIX.

MATZATLI, ſeu Pinea Indica, dicta ſic ab Hiſpanis huius orbis colonis, ob fructum, ſimilitudinem Pineam habentem, planta eſt folia Iridis mittens, ſed ſpinoſa. radicem fibratam, craſſamque. & vnicum pomum, germinibus circumcirca exorientibus, ac altero in culmine Pomi. detractis his ac ſatis, radices ſtatim aguntur, alterumq. producitur pomum Pineis noſtratibus magnopere, vt dictum eſt, ſimile, ſimilibus ſtipatum germinibus. initio hoc apparet rubrum, ſed procedente tempore, ac rubeſcente adhuc germine, palleſcit. Alumna eſt *Haitinæ* Inſulæ, huiuſque nouæ Hiſpaniæ calidarum regionum, montoſiſq. in locis incola. Sapor eſt dulcis, aciduſque. temperies frigida, atque ſicca, ac ſucco manans natura. Exhibetur febrientibus refrigerij gratia, & ad excitandam appetentiam. retineturque ore, vt ſitim mitiget, & linguam humectet. quamquam in bilem facilè conuerti dicatur. Sapor eſt eius fructus, quem vulgus Malocotoneum vocant, non ſine quodam, vt dictum eſt acore. condita aliquando ſaccharo, atque ita delata in Hiſpanias, gratiſſimo

FIGURE 4.4 "Matzatli," the Nahuatl name for pineapple, from Francisco Hernández and Nardo Antonio Recchi (1651), *Rerum medicarum Novae Hispaniae thesaurus, seu, Plantarum animalium mineralium Mexicanorum historia* (Rome: Ex typographeio Vitalis Mascardi, 1651). Courtesy of Getty Research Institute.

Traces of folk classification also surfaced, if fleetingly, in another major plant treatise, *Hortus Malabaricus* (Reede 1678–1703). This unprecedented polyglot work in twelve volumes recorded the uses, descriptions, habit, and habitats of hundreds of plants[16] whose names were recorded in Malayalam, Arabic, Konkani, Portuguese, and Latin (Heniger 1986; Manilal 2003; Raj 2017).[17] As a vector of indigenous plant knowledge, the work's instigator, Hendrik Adrian van Reede tot Drakenstein (1636–91), was, to say the least, unusual. A Dutch East India Company (VOC) commander without the usual medical background, Reede was free of academic preconceptions. In an initiative that was without equal in his time, Reede assembled sixteen local medical and plant specialists, including untouchables and Brahmins, to compile the work. These authorities no doubt employed their own classification of the plants in question. The Dutch editors, however, reorganized *Hortus Malabaricus* following the same Theophrastean categories that had been employed in editing Hernández's treatise on plants in New Spain. According to a major authority on this work, indigenous ordering principles survived where "two, three, four or more plants with similar native names [were grouped] together" (Heniger 1986: 158).[18]

By contrast, in their encounter with East Asian cultures, Europeans were confronted not with unwritten or minimally documented folk knowledge of plants, but rather with long-established traditions of written compilation such as *Shen nong gang mu* (25–220 CE). Li Shizhen's (1518–93) *Bencao Gangmu* initiated a new, idiosyncratic classification (Elman 2007; Métailié and Hsu 2001; Nappi 2009b) that influenced not only Japanese plant investigators, for whom Chinese still functioned as the language of learning, but also Jesuit missionaries such as Michal Boym (1612–59) (Boym 1656). The influence of works such as *Bencao Gangmu* could thereby travel along Jesuit routes of knowledge transmission as had Father Jartoux's climate-based analysis of ginseng's likely habitat in Canada.

Transfer of plant knowledge from the region diminished with China's self-isolation in the middle of the eighteenth century and Japan's restriction of foreigners to the island of Dejima. Engelbert Kaempfer (1651–1716) and Carl Peter Thunberg (1743–1828) of the Dutch East India Company therefore had to conduct their research on Japanese flora while confined to one island. Just like their European contemporaries, however, Japanese officials and men of learning wanted to systematize plant study, while incorporating empirical techniques and foreign knowledge that were increasingly recorded in the vernacular (Ishizu and Valeriani 2015: 57, 61).

SEXUAL DIFFERENCE AND CLASSIFICATION

Sexual difference in plants was already recognized in antiquity. Theophrastus reported that wild plants exhibit male and female characteristics, but warned "we must not assume that in all respects there is complete correspondence between plants and animals" (1916: I.xiv.3–5, 101). His warning was prescient, given that hermaphrodism and asexual reproduction precluded eighteenth-century attempts to derive models of plant sexuality from animal sexuality.[19]

Similarly, the Arabs manually fertilized the date palm by "suspending the male flowers over the female, in order to obtain the fruit" so "[i]t is altogether impossible that they should have been ignorant of a circumstance, which, in these trees at least, is so apparent" (Linnaeus 1786: 2). "Nor is it less certain," Linnaeus concludes, "that the oldest writers have expressly mentioned the sexes of plants. But how little real knowledge of the matter they possessed ... appears from their having frequently described plants as being severally male and female, which were not so" (Linnaeus 1786: 5–6).

Renaissance plant specialists explored the flower's sexual character, although a fuller understanding of reproduction via pollination, including the role of insect pollinators, would take well over two hundred years to crystallize. Andrea Cesalpino (1519–1603), professor of medicine and botany at Pisa, built a systematic classification based on the fruit and seeds, rather than on the leaf, as the ancients had done: "he was the first botanist to recognize and state that the various plant organs ... do not arise haphazardly ... but in definite and regular positions relative to one another" (Morton 1983: 31). Departing from the prevailing pharmacological orientation, Cesalpino argued that "[c]lassification must be based on ... morphological characters of plant organs and parts, not on the properties of the plants in relation to man or any other accidental superficial feature" (Cesalpino 1583, quoted in Morton 1983: 31–2). He thereby contributed to a plant science independent of medicine. Cesalpino's privileging of the flower prepared the ground for Linnaeus' sexual system of classification and its successors (Arber 1986: 168–71).

The English cleric John Ray (1627–1705) laid the foundation for modern classification in *Methodus Plantarum Nova* (1682) (Raven 1947, 1950). Ray divided plants into monocotyledons (one-seed leaf) and dicotyledons (two-seed leaves); this distinction provided a fundamental point of departure for plant classification in the late eighteenth century. Although he enjoys a relatively modest reputation, Ray's insights are of enduring significance; as Rousseau remarked, "*Ray* seems ... to have approached the fundamental method more than anyone else" (Rousseau 2000: 212, emphasis original).

While Cesalpino and Ray expanded botany's morphological basis, Rudolf Jakob Camerer (Camerarius) (1665–1721), professor of medicine and director of the botanical gardens at Tübingen, shed new light on plant sexual difference with his pollination experiments. Camerarius thus gave crucial empirical support to Cesalpino's work and was fundamental for the sexual turn in botany in the eighteenth century. Yet despite its being a ground-breaking study, the *Epistola* has been neglected and only recently translated from Latin (Elvin forthcoming).[20]

Camerarius' insights took hold slowly because major authorities did not accept, much less promulgate, them.[21] Tournefort, whose authority loomed large, held that the reproductive organs perform purely excretory functions: "ignorant of the functions of the sex organs, [Tournefort] disdained the stamens" (Jussieu 1789: xxxi; Stevens 1994: 350). His system likewise ignored the use of the cotyledons as differentiating characters (Greene 1983: 948).

Despite his reluctance to adopt novel concepts, Tournefort remained a key figure well into the eighteenth century, known for his *Elémens de botanique* (1684) (published in Latin as *Institutiones rei herbariae* [1700])[22] and his system of classification, based on the corolla (mono-, poly-, or apetalous), which distributed eight thousand species in twenty-two classes. This system was relatively easy to learn and remained popular, with over twenty major authors adopting it; a well-known French textbook hybridized Tournefort's system with Linnaeus' sexual system (Latourrette and Rozier 1766).

SEXUAL CLASSIFICATION: A CONTESTED LEGACY

On June 10, 1717, Sébastien Vaillant (1669–1722), demonstrator of plants and director of cultivation at the Jardin du Roi (today's Jardin des plantes), Paris, stood in as botany lecturer for Antoine de Jussieu (1686–1758).[23] Vaillant's inaugural lecture was transformative, bringing the sexualist thesis not only to botanists and medical students, but also to the wider public that partook in the Jardin's free courses. The lecture caused a commotion, and Vaillant proved so popular that the students requested he continue lecturing after Jussieu's return (Rousseau 1970: 199, 206).

Vaillant's lecture was published in 1718 and 1727 in a bilingual French and Latin edition.[24] In it Vaillant proclaimed: "*The organs that constitute the different sexes of plants, are principally two. I.e. the* Stamens & *The* Ovaries." Hence, plant parts Tournefort considered "the vilest & and most abject ... are in fact the most noble" (Vaillant 1718: 10, emphasis original). Not surprisingly, these criticisms of Tournefort alienated the Paris scientific establishment, and damaged Vaillant's career (Rousseau 1970: 208–9).

Contemporary hostility notwithstanding, Vaillant's work was not consigned to the dustbin of history. Linnaeus gave it a new lease on life in his *artificial sexual system* of plant classification based on "the number, relative size, and position of the *stamens*, together with the pistils" (Linnaeus 2003: 37, emphasis original). The system is termed "artificial" because it selects a limited set of plant features as the basis for classification. The sexual system thereby "created 24 classes of plants based primarily on the number ... and arrangement ... of [the male] stamens, culminating in *Cryptogamia* (plants without proper flowers). Each class, was in turn, subdivided into orders, usually based on the number and arrangement of female parts [pistils] of the flower" (Jarvis and Cribb 2009: 1). Orders were divided into genera and genera into species. Based on sub-dividing the fructification into thirty-eight parts, Linnaeus concluded that the maximum number of genera would not exceed 5,736 (Linnaeus 2003: 130).[25]

This classificatory construct amounted to more than just a "simple but ingenious arithmetical system founded upon the number of stamens and pistils .... The principal divisions originate in the definitive marks for genera and species, namely, number, shape, proportion, and situation. Absolute number is only one among four of Linné's considerations" (Larson 1971: 58). Shape and situation have, for example, a *spatial* as well as a *numerical* character, forming part of the species description. Relevant shapes include leaves, fruit-body parts, roots, and branches.

The epistemological foundation of the sexual system aligns with Locke's distinction between primary and secondary qualities (Cook 2012: 161).[26] Following Locke, Linnaeus wrote that "[e]very CHARACTERISTIC FEATURE ought to be elicited from number, shape, relative size, and position of all the different parts of the fruit-body" (Linnaeus 2003: 129). He considered secondary qualities to be accidental (with respect to place, time, use, or variable features—color, scent, taste, size, season of flowering) and did not count them as differentiæ (Linnaeus 2003: 129–30).

Jussieu, who was spearheading an important classification system in the 1770s (see below), acknowledged the sexual system's usefulness, but noted the risk of error when examining delicate stamens and pistils using a lens or dissecting needle. He also emphasized that the artificial sexual system selected salient parts in an arbitrary way, resulting in some classes being natural and others not (Jussieu 1789: xxxiii, trans. in Stevens 1994: 352).

These concerns notwithstanding, the artificial sexual system proved its value in combatting exposure to too much information, eclipsing the many classification systems available at the time (Adanson [1763–4] 1966: lxxxviii, iv–v, lxxxiii–xciii). Plant explorers, in particular, adopted Linnaeus' system because of its ready applicability as "a classification that the human mind could encompass" (Stevens 2002: 15; see also Linnaeus 2003: 111).[27] Linnaeus, himself a keen cameralist[28] and acclimatization

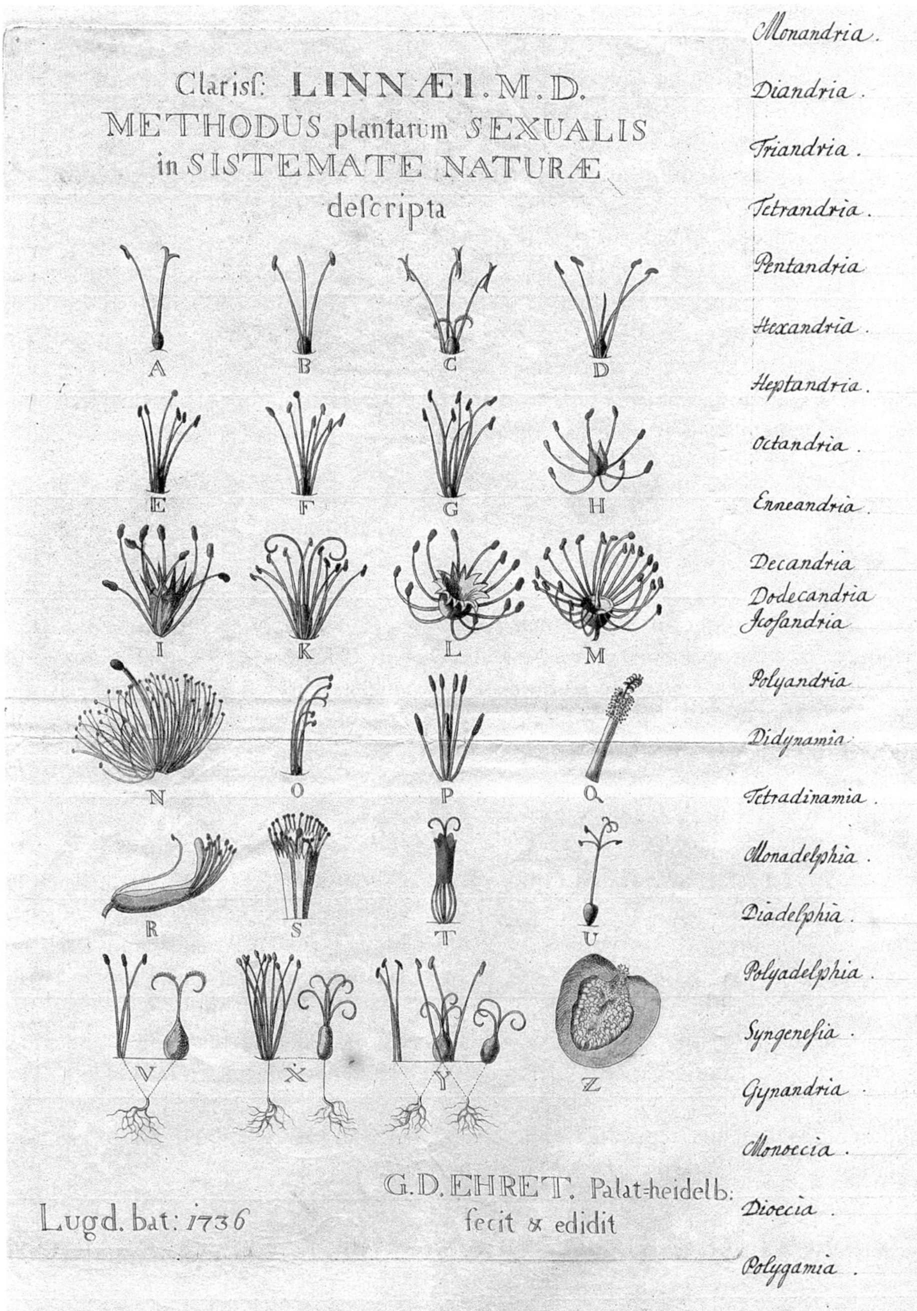

FIGURE 4.5 Diagram of the artificial sexual system executed by Georg Dionysius Ehret (1736), *Clariss: Linnæi. M. D. Methodus plantarum sexualis in sistemate naturæ descripta* (Leiden). Courtesy of Uppsala University.

advocate, thus provided crucial information infrastructure for overseas projects of prospecting, trade, and empire (Batsaki et al. 2016; Cook 2001; Koerner 1999). With such a useful cataloging system to hand, traveling botanists—most famously Joseph Banks (1743–1820) and Daniel Solander (1733–82) on James Cook's first *Endeavour* voyage—were equipped as "agents of empire" (Mackay 1996).

By the late eighteenth century, the artificial sexual system was falling by the wayside as specialists adopted the natural family classification inspired by Bernard and Antoine-Laurent de Jussieu; Linnaeus himself endorsed this development, regarding the sexual system as only a way-station (Linnaeus 2003: 40). The Jussieus relied not just on the reproductive organs, but also on overall appearance (*habit*) and weightings of various plant features. The natural family system, which drew on Ray's work in the preceding century, formed the basis for important taxonomic work by, among others, the Genevan Augustin Pyramus de Candolle (1778–1841). As the book of nature's "table of contents," the sexual system "was a failure" (Stevens 1994: 267), which in 1836 was declared dead (Lindley 1836: vii).

LINNAEUS' LEGACY AND ITS CRITICS

In contrast with other defunct classification systems, Linnaeus' artificial system of sexual classification is unique in that its manifold implications continue to be debated. The critiques have ranged from puritanical diatribes to contemporary feminist history of science. Johann Georg Siegesbeck (1686–1755) of the St. Petersburg Academy of Sciences condemned the sexual system as immoral; the matter did not end there (Sokoloff et al. 2002: 150, 158–61) since Linnaeus took revenge by naming St. Paul's Wort, a smelly plant that grows in the mud, *Sigesbeckia orientalis* (Rowell 1980: 16, 19–20). Ironically, the same St. Petersburg Academy awarded a prize to Linnaeus' *Dissertation on the sexes of plants* (Linnaeus [1761] 1786).

Like Siegesbeck, more recent critics object to how Linnaeus incorporated gender into the sexual system. They cite his elaborate human-plant analogies (explicitly characterized flower parts with reference to human sexual organs) and conjugal imagery: "the CALYX is the *bedroom*, the COROLLA is the curtain" (Bewell 1996: 174–5; Linnaeus 2003: 105, emphasis original). While some feminist scholars argue that Linnaeus based the sexual system on contemporary gender hierarchy,[29] this view is disputed by leading experts (Drouin 2008: 189; Müller-Wille 2007: 548). The fascination with plant sexuality lives on, as evidenced by a recent collection of lavish photographs of plants' reproductive organs (Kesseler and Harley 2004), not to mention the photographic work of Robert Mapplethorpe (1946–89).

COLLECTING AS DATA MANAGEMENT: THE HERBARIUM

> The herbarium is better than any picture and necessary for every botanist ...
>
> (Linnaeus 2003: 18)

Leonhard Rauwolf, who was discussed above prospecting for plants in the Near East, returned to Augsburg in 1576 with two hundred dried plant specimens, that he added to his *herbarium*, a form of data management in which carefully dried and preserved plants were attached to pages bound into books. While he was not the first to constitute such a collection, Rauwolf was the first to incorporate plants from a vast, largely unknown area

FIGURE 4.6 *Vitex agnus-castus*, collected in the Near East by Leonhard Rauwolf (1535–96), Herbarium of the Naturalis Biodiversity Center, Leiden—Registration number L.2111378.

into the new format.[30] His elegant herbarium, preserved in Leiden, is one of the oldest and most precious in existence (Ghorbani et al. 2018).

The precise origins of the herbarium are unknown; while usually considered a sixteenth-century invention, collections of dried plants may have been compiled earlier by, for example, the Andalusian physician, Ibn al-Bhaytar (*c.* 1190/7–1258) (Cabo

Gonzalez 1997: 25).[31] By the sixteenth century the greater availability of paper due to the invention of printing, together with botany's intimate connection with medical education, contributed to the herbarium's rapid rise in popularity (Benkert 2016; Boutroue 2008; Saint-Lager 1885).

In the herbarium the fragile, short-lived plant was transformed into a long-lasting, physical record incorporating not only the dried specimen, but also combinations of the following: plant name, collection date and site, name(s) of collector(s), depictions of missing plant parts, and depictions of environments, as in herbaria made by Hieronymous Harder (1523–1607) and Jean-Baptiste François Rozier (1734–93).[32] Specimens might also be aligned for comparison next to plant illustrations, as in Felix Platter's (1536–1614) herbarium (Benkert 2016).

Dried specimens, which mutely testified to their myriad origins, uses, and kinship with already-known plants, could be organized in various ways, from chronological to systematic. A systemic arrangement following Tournefort or Linnaeus, for example, facilitated later retrieval of specimens for consultation and comparison with other plants or specimens.

Innovations during our period made this form of data management more flexible. For example, Rousseau experimented with miniaturized formats to use in the field, give to friends, and popularize botany (Cook 2012: 290). Often exhibiting great aesthetic as well as scientific value, herbaria such as Harder's might do double duty as prestige objects gracing the collections of royalty and well-to-do savants.

FIGURE 4.7a and b Herbarium cabinet of Linnaeus before and after restoration. © Linnean Society of London.

As a paper technology for plant data management, the herbarium offered many advantages. It acted as a "winter garden [hortos hyemales]" available for study when other options were unavailable (Tournefort 1694: 547; van den Spiegel 1606: 79). It also had epistemic value (Müller-Wille 2017). By displaying morphologically similar plants together, the herbarium facilitated comparison among specimens, between specimens and illustrations.[33] Cabinet storage in a loose format facilitated taxonomic revision, such as the transfer of species between genera (Lamarck 1789). Linnaeus designed his own cabinets, while smaller collections might have been stored in boxes, the option adopted by Rousseau. Finally, the herbarium surpassed the physical capacity and seasonality of botanical gardens with an *unlimited number* of plants (Bulliard 1783; Lamarck 1789).

The role of physical collections in botanical work was already manifest by the early seventeenth century: for example, Bauhin's *Pinax* directly reflects his herbarium collection, which, like modern ones, was unbound and could be reordered as needed (Benkert 2016). The herbarium as plant register, teaching tool, and reference work therefore soon proved indispensable to all manner of plant specialists, from apothecaries to professors of medicine, from ship's surgeons to territorial rulers (Cooper 2018).

The core collections of today's herbaria housed in botanical gardens and universities attest to their hybrid origins in private collections that incorporated specimens acquired from commissioned collectors, missionaries, traders, laypersons, and ship's surgeons, among others. It was common practice for collectors to gather two or more specimens of a particular species, keeping one and presenting or exchanging the others for plant specimens or other natural history items, including books. The specimens thus entering into this system of circulation sometimes have complicated histories and autobiographies, usefully characterized by the term "silent messengers" (Dupré and Lüthy 2011).

Exemplary of such silent messengers are specimens acquired by Sir Hans Sloane (1660–1753), that he bequeathed to the British nation with his other diverse holdings to form the core of the British Museum.[34] A wealthy natural history amateur, Sloane not only collected plants himself while serving as physician to the Duke of Albermarle, governor of Jamaica, but also acquired herbaria compiled, most notably, by London apothecary James Petiver (*c.* 1665–1718). Petiver's collection in turn contained specimens collected in the Philippines by Jesuit apothecary Josef Kamel (1661–1706) and in the Iberian peninsula by Barcelona apothecary Joan Salvador I Riera (1683–1726).[35]

To order these collections, their creators employed their classification systems of choice, or simply mounted their specimens chronologically in a working herbarium. They employed different techniques when fastening the specimens to the page with varying degrees of impact on the specimen's physical integrity: attaching the specimen with glue or wax was common, but harmful; careful practitioners such as Rousseau used small strips of paper glued to the paper on either side of the specimen. This approach is still favored today alongside other methods such as sewing. These transformations of technique rendered the herbarium an essential reference work and teaching tool (Boom 1996; Boone and Vivent 2013; Bridson 1992; Nesbitt 2014).

As part of private museums—*cabinets d'histoire naturelle*, *Naturalienkammer*—herbaria joined national natural history collections in the nineteenth and twentieth centuries, becoming irreplaceable repositories of the planet's plant heritage, including rare and extinct species. The largest herbarium in the world is housed in the Jardin des plantes, Paris, and contains approximately nine million specimens. Of these, 6,963 come from Tournefort, who not only collected in Europe, but also in the Near East. The collections of Tournefort and others originally belonged to the French King's natural history collection ("Cabinet du roi"), neither clearly private nor public until appropriated by the French republic during the Revolution.

The five million herbarium specimens of the Conservatory and Botanical Garden of Geneva, Switzerland present a similar trajectory, albeit without royal involvement. The core collection of 250,000 specimens was assembled by Franco-Swiss banker, Benjamin Delessert (1773–1847), inspired by his mother's botanical correspondence with Rousseau (Rousseau 2000). Delessert acquired specimens from collections made by Linnaeus, Amsterdam professors, Johannes (1707–80) and Nicolaas Burman (1734–93), and explorers Philibert Commerson (1727–73), Jacques-Julien Houtou de la Labilladière (1755–1834), and André Michaux (1746–1803). While nothing of comparable scale arose outside Europe, collections of botanical specimens are attested in eighteenth-century Japan (Ishizu and Valeriani 2015: 61). As with botanic gardens, imperial powers established herbaria to inventory, and thus control, colonial resources, with the earliest founded by the British in India: Kolkata (1786) and Howrah (1795).

The modern herbarium holds type specimens that anchor the plant's binomial name to a particular specimen. In case of doubt about a plant's identity, the type specimen supplies the definitive reference material, thereby eliminating—in theory, at least—referential ambiguity. Herbaria, therefore, play an essential role in stabilizing nomenclature and classification, necessary not only for botany, but for medicine and other sciences that use and study plants. Digitization initiatives have further enhanced the value of the herbarium by making type specimens available electronically.

Modern anti-insect measures and temperature-humidity control enable herbaria to serve as databases of long-lasting physical material that constitute an indispensable record of biodiversity, rare and extinct species, and their DNA. Specimens may even contain potent alkaloids and other chemicals, key targets of drug research (Nesbitt 2014).[36] Thus, seventeenth- and eighteenth-century herbaria may as yet yield information useful in addressing urgent problems confronting the human species and the planet.[37]

"PHYSICISTS" INVESTIGATE "LAWS OF VEGETATION"

While botanists studied plants' exterior parts, physicists, following the Royal Society's motto, "nullius in verba" (rely on no one else's word), investigated "laws of vegetation" (Latourrette and Rozier 1766). Their experimental methods ranged from the anatomical to the chemical and from the horticultural to trial and error. Research subjects—might include oneself—widely used due to the dearth of research subjects, other human beings, or animals.

As with synonyms, two major imperatives drove plant experimentation: health concerns and commerce. One plant that answered both of these imperatives was tea, at once both luxury beverage and pharmaceutical; in its various guises as green and bohea tea, it was subjected to any number of experimentation techniques, starting with transplantation of the plant itself (discussed below), coupled with chemical procedures carried out on tea leaves. In the seventeenth century, distillation was widely used, but generally produced the same substances, irrespective of the plant in question; by the early eighteenth century, experts therefore concluded that distillation failed to yield new medicinal substances and should be replaced by other processes (Short 1730: 17). Physicians such as Johann Pechlin (1646–1706), Thomas Short (*c.* 1690–1772), and John Coakley Lettsom (1744–1815) tested tea infusions with a variety of chemicals, but Short and Lettsom sought more direct evidence of tea's medicinal and/or toxic effects on the human body. Unusually for his era, Short was mindful not to cause "Danger to any animal," applying infused tea only to mutton. Lettsom, on the other hand, sought to test the validity of Chinese avoidance of green tea less than a year old by injecting a strong infusion of young tea directly into live frogs (one of whom died as a result); he thereby considered the Chinese custom validated.

By the early eighteenth century, popular works in the vernacular stressed their reliance on rigorous, yet *accessible* experimentation that could be practiced by anyone (Short 1730: 17). Given the artisanal and informal turn in contemporary history of science, it is important to touch briefly on the democratization of experiment spearheaded by Short and others. Indeed, throughout human history crucial experimental work on plants and their products has been taking place in informal, familial, and artisanal settings the world over. Consider Amerindians' discovery of "cooking maize grains in a lime solution and then soaking them to obtain nixtamal before grinding the maize to make dough or masa. This process liberated the bound niacin, making it bioavailable for digestion" (Eamon 2018: 150), thus preventing nutritional deficiency. When Europeans adopted maize in their diet, they neglected the crucial addition of lime, thus falling victim to pellagra, first described in Asturias (Spain) in the eighteenth century. Research into Early Modern household medicine production reveals crucial health-related experimental work with plants and their products in familial, non-elite settings (Leong 2008). This was not heroic science, but work with and on plants carried out by persons who would never make their appearance in a learned academy, including women, slaves, and people who could not read. These were not "physicists" seeking the "laws of vegetation," but ordinary people intent on improving the material conditions of their lives, their diets, and/or their health.

Short appealed to this broader, non-elite audience by claiming his "Experiments are easy, and practicable by every curious Person on any Plant, without Experience, much Apparatus, Loss of Time, Danger to any Animal, or Acquaintance with the chymical Jargon of Words, more like Conjuration than Instruction" (Short 1730: 17). Lettsom followed Short in providing clear explanations of repeatable experiments. Short and Lettsom thereby helped to make their approach to experimentation accessible to those who were excluded from collectives of learned gentlemen where experiments were witnessed and subjected to canons of verification.

Like artisanal practices of herbalists and women in households, transplantation—surely a very ancient form of experiment—should be considered on a par with experimentation by learned physicians and savants.[38] (Note that natural transplantation likewise occurs by seeds being carried by birds, other animals, or wind.) However, transplantation as cameralist state policy, supported by state agencies and resources, was for the first time being practiced on an unprecedented scale. Plants of particular interest included trees for naval shipbuilding, spices, coffee, and tea—due to their high cost, quinine due to its critical role in combatting fevers, and citrus fruits, particularly prized by European elites.[39]

Acclimatization was no simple matter; if live plants were being transported, the first challenge was how to keep them alive on months-long sea journeys to Europe or other locations. Since sea water is totally unsuitable for this purpose, fresh water had to be stored on board, a serious burden for vessels already at their limit to carry water and provisions for their crews. Various methods were tried in order to keep plants alive during long sea voyages; Captain Bligh's cabin on the notorious HMS *Bounty* (1787–9) was converted under the supervision of Royal Society President Joseph Banks into a floating breadfruit hothouse equipped with ample natural light and special water drainage.[40]

Some transfers were more covert than *Bounty*'s mission or Linnaeus' long struggle to obtain and grow tea in Uppsala (Hodacs 2016: 153). In a daring mission, Pierre Poivre (1719–86), a former Jesuit missionary in China, stole extremely valuable clove and nutmeg plants from the tightly controlled Dutch East Indies and introduced them in the French colonies of Île de France (Mauritius) and Île Bourbon (Réunion)—the feat that clinched his career as a colonial administrator.

Even if foreign plants survived long sea voyages, changes in soil, temperature, humidity, and other factors posed huge challenges to acclimatizing them in Europe or in European colonies. Special care and infrastructure were required, with plants wintering in purpose-built heated glasshouses pioneered by the Dutch in the seventeenth century; even then they might only survive for a brief period, like Linnaeus' tea plant in the Uppsala botanic garden. Successful transplants gained renown and would be treated like celebrities, visited by the curious and the learned.

One of the eighteenth century's acclimatization masters, Claude Richard (1705–84), presided over *c.* four thousand species of rare plants in the Trianon garden at Versailles, the largest collection of its kind in France (later replaced with an Anglo-Chinese landscape garden by Marie Antoinette). For Linnaeus Richard was "the ablest horticulturalist that Europe has ever known" (Choffé 2004: 57). Under the patronage of King Louis XV, Trianon became known as a "botanical laboratory," not only for successful acclimatizations such as the Virginia tulip poplar tree (*Liriodendron tulipifera*), but also as the site where Bernard de Jussieu's natural family system, precursor to Antoine-Laurent de Jussieu's natural method of classification, was first applied in 1759 (Choffé 2004: 57; Jussieu 1789).

As explained above, some transfers, like that of maize (corn), badly misfired. Entire civilizations consumed corn for centuries without developing nutritional deficiency. The European adoption of maize demonstrated not only the potential pitfalls of plant transfers, but also the importance of indigenous knowledge, together with the oldest of all experimental methods: trial and error. By contrast, European adoption of tea showed more awareness of the value of indigenous knowledge: the Dutch conserved the Chinese habit of drinking tea after meals and never on an empty stomach, while some Europeans—to their detriment— disregarded traditional use of tea in its place of origin, drinking it first thing in the day (Garcin 1742). Lettsom, too, was convinced the Chinese knew the most beneficial way to take their tea, confirming the validity of this custom at the expense of the unfortunate frog injected with strong, young tea.

No discussion of plant experimentation in this era can be complete without reference to experiments that followed on William Harvey's (1578–1657) discovery of the circulation of the blood such as the discovery by Marcello Malpighi (1628–94) and Nehemiah Grew (1641–1712) of plant vascular systems. Striking experimental breakthroughs were also achieved by Stephen Hales (1677–1761), and Jan Ingen-Housz F.R.S. (1730–99), whose work on the effect of light on plants led to the discovery of photosynthesis. In 1735 Hales' work gained international traction with the publication of his *Vegetable Staticks* (1727) in a French translation by no less than Georges Louis Leclerc (1707–88), comte de Buffon, later *intendant* of the Cabinet du roi, and a renowned natural history authority in his own right.

CONCLUSION

This brief survey has shown how nomenclature, classification, and collections of physical specimens served as crucial data management technologies during the seventeenth and eighteenth centuries. These technologies in turn opened the way to new epistemic insights and transformations of knowledge.

It is important to stress that the debates and developments reviewed here were not solely elite or specialist concerns; rather, broad segments of the literate European public informally participated in what is now known as "citizen science," managing plant data, making empirical observations, collecting, naming, and classifying plants. These activities differed both in origin and practice from the mathematical and mechanical work characteristic of what is taken to constitute the classic "scientific revolution," itself an increasingly contested idea (Eamon 2018; Raj 2017).[41] Indeed, the tools for tackling the Early Modern information explosion extended from scholastic philosophy (Bauhin's definitions) to a vast range of traditional knowledge systems and networks worldwide (Raj 2016).

Counter to standard accounts, it has been argued here that Europe was hardly the sole venue for collecting, naming, and classifying endeavors. In fact, European initiatives in botanical nomenclature, classification, and specimen collection depended on largely unacknowledged, subaltern sources of natural knowledge such as slaves and women healers, who supplied crucial contributions to the global project of plant knowledge spearheaded by Europeans.

Subaltern contributions surface where we might least expect to find them: in the work of Linnaeus, despite his pronouncements against contaminating taxonomy with pharmaceutical allusions, "barbaric" names, and references to local uses. All of these nonetheless infiltrated binomial nomenclature, classification, and herbaria, where they remain to this day (Cook 2010; Manilal et al. 2003). Yet these subaltern contributions to modern scientific nomenclature have been largely ignored in condemnations of "linguistic imperialism" and calls for the decolonization of scientific names (Needham 1986; Schiebinger 2004b). In view of such knowledge erasures, it is time to acknowledge myriad Latinized words from Arabic, Persian, Sanskrit, Malayalam, and Caribbean languages in modern plant nomenclature.[42] Let us recall these cultures' presence in scientific names when drinking our *Coffea*, eating our *Spinacia*, and washing with our *Sapindus*!

CHAPTER FIVE

Plants and Medicine

CLARE GRIFFIN

In 1645, Tsar Mikhail Fedorovich (b. 1596), the first of the Romanov Dynasty to rule Russia, lay dying. He was examined by his German doctors, and prescribed a series of compound medicines containing several plants, among them: rhubarb (*Rheum* spp.), transported from China overland to Moscow by Central Asian merchants; senna (*Senna alexandrina*), cultivated in North Africa and brought to Russia via western Europe; juniper berries (*Juniperus communis*), which grew in the traditional Russian lands surrounding the capital and were collected by local peasants as part of their service to the tsar; and sassafras (*Sassafras albidum*), a tree growing in Florida and other more southerly parts of the North American continent, collected by Native Americans and shipped across the Atlantic by the Spanish (Mamonov 1881: 120). All of this is recorded in the files of the seventeenth-century Russian Palace medical department, the Apothecary Chancery, in some detail (Levin 2004). After all, the death of the tsar is not only the death of a man, but a shock for the political system, no less in this case as Mikhail Fedorovich's son and heir, Alexei Fedorovich, was all of sixteen at the time. The prescriptions for Mikhail Fedorovich were as much a political maneuver to retain a strong and mature ruler as they were a medical case. This medical attempt to fix a political problem was unsuccessful, and Mikhail Fedorovich's death caused much trouble to the recently established Romanov Dynasty. What concerns us here is not the failure of the attempt, but rather the methods employed. What does the use of plants from as far east as China and as far west as Florida in the Moscow court in the middle of the seventeenth century tell us about plants and medicine during the Enlightenment?

The most obvious point here is about the geography of the Early Modern global world, and the place of Europe—which is the nexus of this volume—within it (Parker 2010). There have now been many debates about Early Modern globalization, focused particularly around the inception and developments of American-Eurasian links framed around the notable dates of 1492—Christopher Columbus' (1451–1506) Spanish-sponsored voyage to the Americas—and 1571—the establishment in the Philippines of the port of Manila, by the Spanish, facilitating American-East Asian trade—as well as the strengthening of interactions across Eurasia through the creation of the Littoral System of costal navigation between East Asia and Europe via southern Africa, and the continued importance of Eurasian land routes like the famed Silk Roads (Breen 2012; Flynn and Giraldez 2006; Frankopan 2015; Riello 2009). This was not a completely interconnected world. That other major populated land mass, Australia, would not begin to be directly integrated into intercontinental exchanges until the very late eighteenth century, after James Cook's British-sponsored voyage of 1770, although the Dutch were already traveling there in 1606. Nor was it like the modern globalized world: commodity price convergence, where the prices of the same good sold in different locations around the globe began to come closer together; increased speeds of travel and goods transportation through innovations like the steam-ship and the railway; and communications technologies, such as the telegraph lines laid around the globe, were all issues of the nineteenth century and later (Heng 2014; Pieterse 2012). Early Modern globalization was rather characterized by increasingly close links across Afro-Eurasia, and between Afro-Eurasia and the Americas, links commonly driven by a desire for commodities, including medicinal plants. By the middle of the seventeenth century, high-status patients like Mikhail Fedorovich were able to exploit such connections to consume medical drugs bringing together plants from across the known world. During the next century and a half, high-status patients would continue to desire and consume such medicines, and more humble sufferers would also have increasing access to them. The plants of Enlightenment medicine were global.

A second major point here is chronology. Rhubarb had been traded across Central Asia for centuries; sassafras was only known to Eurasians after 1492. The original prescriptions for Mikhail Fedorovich were written in Latin, as was common practice at the Early Modern Russian court, and indeed across Europe. The use of the ancient world language of Latin in prescriptions, and in a great number of other Early Modern medical texts, was an artifact of a broader concern with the past. European intellectuals of all stripes saw themselves as living in a long tradition of thought and practice, notably connected back to the Greco-Roman ancient world, and the major medical writers of Hippocrates, Galen, and—most importantly from the point of view of medical drug practices—Dioscorides. Much of what was practiced at the Early Modern Russian court was deeply rooted in the Early Modern world's ideas about ancient medicine: both rhubarb and juniper berries can be found in Dioscorides' *De materia medica* (*c.* 50–70 CE) (Beck 2005). Elsewhere in the Early Modern world, Chinese medicine and Ayruvedic medicine similarly drew upon contemporary editions of ancient texts to inform their herbal medical practices. These ancient texts were always limited in the geography of the plants they included. Sassafras, a tree in the mid-seventeenth century only known to grow in Florida and surrounding regions of North America, was not found in Dioscorides, nor in the works of any other pre-contact Afro-Eurasian medical writer. That Floridian tree, along with a number of other American plants, continued to be important for the next 150 years. In Enlightenment Eurasia sassafras and other American plants were being prescribed to princes, sold on markets, and written about in self-help texts aimed at the lowliest of Early Modern medical readers from Madrid to Nagasaki; during the same time period those plants continued to be used in the Americas by Native Americans, European invaders, and enslaved Africans. Enlightenment medicinal plants thus solidly belonged to the globalizing Early Modern world, showing both long-term continuity and substantial novelty.

THE IMAGINED PAST OF ENLIGHTENMENT MEDICINE

People in the Enlightenment saw themselves in terms of the past they thought had existed (Shaw 1989). Concerning medicine, this meant ideas about pre-modern medical texts, practices, and products, and the supposed relationship between those works and a line of imagined continuity through time and space to the Enlightenment; Mikhail Fedorovich's doctors in 1645 were writing in an ancient world language, and drew upon Early Modern versions of ancient world medical ideas, medicaments, and texts, in constructing his treatment. That line of continuity was increasingly interrogated in Early Modern Europe, with the humanists, and medical humanists, deconstructing the supposed linkages, and revealing the much more complex transformations of medical ideas through centuries of translating and transcribing. In the sixteenth century, the medical humanists devoted attention to the medicament "mummia," a powder created from powdered human corpses. Tracing the lines of translation and transmission, they established that the use of such corpse powder was a misapprehension of the meaning of earlier texts, which recommended a substance used in the preparation of Egyptian mummies, not the preserved bodies themselves (Dannenfeldt 1985; Siraisi 2007: 229–30). However, the imagined link was never truly severed. Long past the zenith of Renaissance humanism—the sixteenth century—ancient world texts continued to be used; mummia continued to be prescribed in Europe at least into the eighteenth century. Indeed, perhaps the most significant achievement of the humanists was to strengthen the imagined continuity, by proposing that the problems of transmission were both fixable and fixed. In terms of plants and medicine, what loomed large in this imagined past?

Dioscorides was a name familiar to all Early Modern medical men in Europe and North Africa, an ancient world thinker (*c.* 40–90 CE) famous for his Greek-language work on medical drugs, conventionally referred to by its Latin title, *De materia medica*, on the materials of medical treatment, i.e. medical drugs. Dioscorides was the ultimate source of authority on *materia medica*, plants included. The other two major ancient authorities relied upon by Enlightenment Europeans were Hippocrates (*c.* 460–*c.* 370 BCE), a Greek physician, and Galen (129–*c.* 200 CE), a prominent Greek physician working in the Roman Empire. The balancing of the four humors—blood, phlegm, black bile, and yellow bile, whose correct proportions in the human body was held to be the key to a healthy life—a theory advocated by Hippocrates, expanded by Galen, and reworked and copied for centuries thereafter, could be achieved in a number of ways. One famous method was venesection, bleeding a patient of "excess" blood. But medicines, including plant-based medicines, were also commonly used to rebalance the humors. These ancient world texts were all repeatedly reworked across the centuries: Dioscorides strongly influenced the medieval European herbal tradition, collections of useful knowledge on natural objects, arranged by object, predominantly, but not exclusively, plants (Collins 2000). The tradition of the herbal, based on Dioscorides, encoded and protected the practice of plants as medicines, creating texts that would later be used by Enlightenment practitioners.

Just as Early Modern Europeans drew on contemporary reworkings of much older texts for their medicines, people living elsewhere in the Early Modern global world did similarly. Two other major textual traditions are Chinese medicine and Ayurveda. Chinese medicine relied, and continues to rely, upon modern versions of ancient texts, such as the *Shanghan Lun*, or *Treatise on Cold Damage*, a medical treatise including a number of herbal remedies composed by Zhang Zhongjing (*c.* 150–*c.* 220 CE) sometime before 220 CE. Chinese texts on various topics had a wide geographical spread, being influential not only within the Empire, but also east into the Korean kingdoms and Japan (Scheid 2013). Ayurveda is a South Asian medicinal tradition drawing upon texts covering everything from surgery to herbal medicine that were composed around the sixth century BCE and were reworked several times across the centuries (Wujastyk 2003). In all these cases, medical practitioners relied—or continue to rely—upon revised editions of ancient works, which have both changed across the centuries but also represent an important textual continuity.

Alongside these traceable textual traditions, much of pre-modern, and Enlightenment, healing practices was not recorded in texts, but rather circulated in oral tradition, making it much harder for historians to appreciate and evaluate. Quinine is derived from the South American plant cinchona (mostly *Cinchona calisaya*), and has been used to treat fever, in particular malarial fever. Its appearance in texts can be dated to the Early Modern period. Yet its actual period of usage is likely much longer. Quinine was heavily used in Enlightenment European medicine, but knowledge of the plant's medicinal applications was derived from local Andean healers, the Curanderos, who were active in the region of modern-day Ecuador and Peru. Archaeological and anthropological evidence suggests that the healing practices of this group substantially predate the Columbian period (Crawford 2016: 26). There are similar suggestions of long-standing healing practices, commonly involving herbs, that were preserved orally rather than textually, worldwide. Medical texts provide vital evidence for the historian, but can only ever provide an incomplete picture of healing.

The imagined past of plant medicines in the Enlightenment stretched back to the ancient world, drawing on very old texts that underwent substantial changes across the centuries, and also long-standing—if hard to trace—oral traditions. This past cemented for Enlightenment medical practitioners the centrality of plants to medicine.

OLD WORLD TRADE AND MEDICINAL PLANTS

The imagined textual past of Eurasian Enlightenment medicine had distinct limits; so, too, origins of plants in those texts were circumscribed. Mikhail Fedorovich's 1645 treatment included one American plant, but several from the Old World; the former was a fairly new development. Before 1492, plants used in Eurasian medicine were exclusively sourced from Afro-Eurasia or, to give a less modernizing and cumbersome term, the Old World. Europe had long had strong connections to the Middle East and North Africa; the relatively small and calm Mediterranean long served as more of a facilitator than a barrier to trade and contact. From the Mediterranean, one could travel east via the various land and littoral paths making up the Silk Roads, as Marco Polo (1254–1324) most famously did in 1271, following long-established and much-trodden land routes. From the sixteenth century onward, following the Portuguese explorer Vasco de Gama's (1460s–1524) successful sea voyage from Europe to India in 1497–9, European merchants began sailing more of the route to East Asia, joining up older regional passages into what became known as the Littoral System, following the coast of Africa. This new(ish) route did not displace the Silk Roads, winding across the Central Asian landmass, but rather they both acted to increase trade between the two ends of Eurasia, making medicinal plants increasingly accessible far from their region of cultivation (Levi 1999).

Europe was part of a geographical and botanical continuity, in particular linking it to the rest of the Mediterranean world. Indeed, a number of plants mentioned in Enlightenment and earlier European medical texts could have been sourced from across that broader region. One such plant was senna. *Senna* species, legumes that grow across the Old World tropics and in parts of the Eurasian temperate zone, were for centuries in use as a medicine. The most famous application was of the pods as a laxative, a use that continues to the present day, but their properties as a gum also made them important as a non-active ingredient in certain medicines, notably plasters, to help them stick to the skin. The senna prescribed to Tsar Mikhail Fedorovich in 1645 was described in the Russian-language report on his illness as *aleksandreiskii list*, suggesting a link to Alexandrine senna

(*S. alexandrina*), a reference to the ancient Egyptian port city, and so pointing to a North African origin for Russian senna supplies (Mamonov 1881: 120). Here, we cannot be sure of the exact geographic origins of this plant, demonstrating the fundamental botanical connections of Europe to its neighboring regions.

More of an obvious botanical disjunction is presented by rhubarb. Modern culinary rhubarb (*Rheum rhaponticum* and *R. rhabarbarum*) are different species from those sought by Early Modern Europeans; of all the foreign plants imported to Europe in this period, unlike culinary rhubarb, medicinal rhubarb was one of the most desirable, and also one of the hardest to transplant. The very name tells Greek speakers of its extra-European origins, a plant from beyond the Rha (the present-day Volga river in European Russia). Early modern practitioners most highly valued Chinese rhubarb (*Rheum palmatum*), and significant effort and money was put into importing it to Europe from East Asia. Medicinal rhubarb was one of a number of living plants thought to travel poorly by sea; the Bukharans—a trading nation based out of the ancient Silk Roads market city of Bukhara, in present-day Uzbekistan—and the Russians both made good profit as middle-men facilitating its trade across Central Asia, and its sale to western Europeans (Burton 1997; Foust 2014; Romaniello 2016: 19). Indeed, rhubarb was of such value that the Russian state went looking for a source within its own borders in the mid-seventeenth century, sending explorers to the Siberian territories in an ultimately unsuccessful attempt to put the medicinal rhubarb trade more firmly under their control (Monahan 2011, 2013). More generally, Central Asian trade was a major factor in worldwide exchange in this period, with various middle-men—Armenian and Indian merchants as well as the Bukharans and the Russians—facilitating (and profiting from) the trade between the two ends of the Eurasian continent (Aslanian 2011; Dale 2002; Levi 2002). The difficult nature of the medicinal rhubarb trade reveals its value to Enlightenment Europe: across this period, an expensive and fragile plant, tricky to transport, was consistently sought and used.

Medicinal rhubarb was often transported from one end of Eurasia to the other by land; other South and East Asian commodities popular in Europe were more commonly moved by the sea route. One such plant was nutmeg (*Myristica fragrans*). Sourced from what were once called by Europeans the Spice Islands, the Maluku islands in present-day Indonesia, nutmeg was a hugely valuable culinary and medicinal commodity in the Early Modern world. The Spice Islands were dominated by the Dutch, and their trading company, the Vereenigde Oost-Indische Compagnie, or Dutch East India Company; in the middle of the eighteenth century, the French managed to successfully transplant nutmeg plants to their colony, the Île de France (present-day Mauritius), an effort led by the Compagnie des Indes, the French East India Company (Spary 2005). In both cases, those companies used the Littoral System, winding its way from Southeast Asia, around the coasts of India and Africa, all the way west and north to Europe, to transport the priceless seeds to European markets. Here, nutmeg could then travel further, being resold across the continent. In the 1660s, the Moscow court commonly purchased it from the north German port city of Hamburg, then a major trading center as a part of the Hanseatic League, a medieval trading monopoly that long dominated the Baltic Sea trade (Griffin 2017). This nutmeg most probably entered Europe via the Littoral System. Taking nutmeg from Southeast Asia to Moscow via western Europe is a geographically eccentric route. Nor was it the only possible one: spices such as nutmeg were also on sale at the regular market on Lake Yamysh, on the Irtysh river in present-day Kazakhstan, a market from which Moscow merchants commonly sourced goods (Monahan 2016: 192). Having

FIGURE 5.1 Rhubarb, as depicted in Pietro Andrea Mattioli's Italian translation of Dioscorides, Venice, 1568. Wellcome Images (https://wellcomecollection.org/works/wzkk96pv). Attribution 4.0 International (CC BY 4.0).

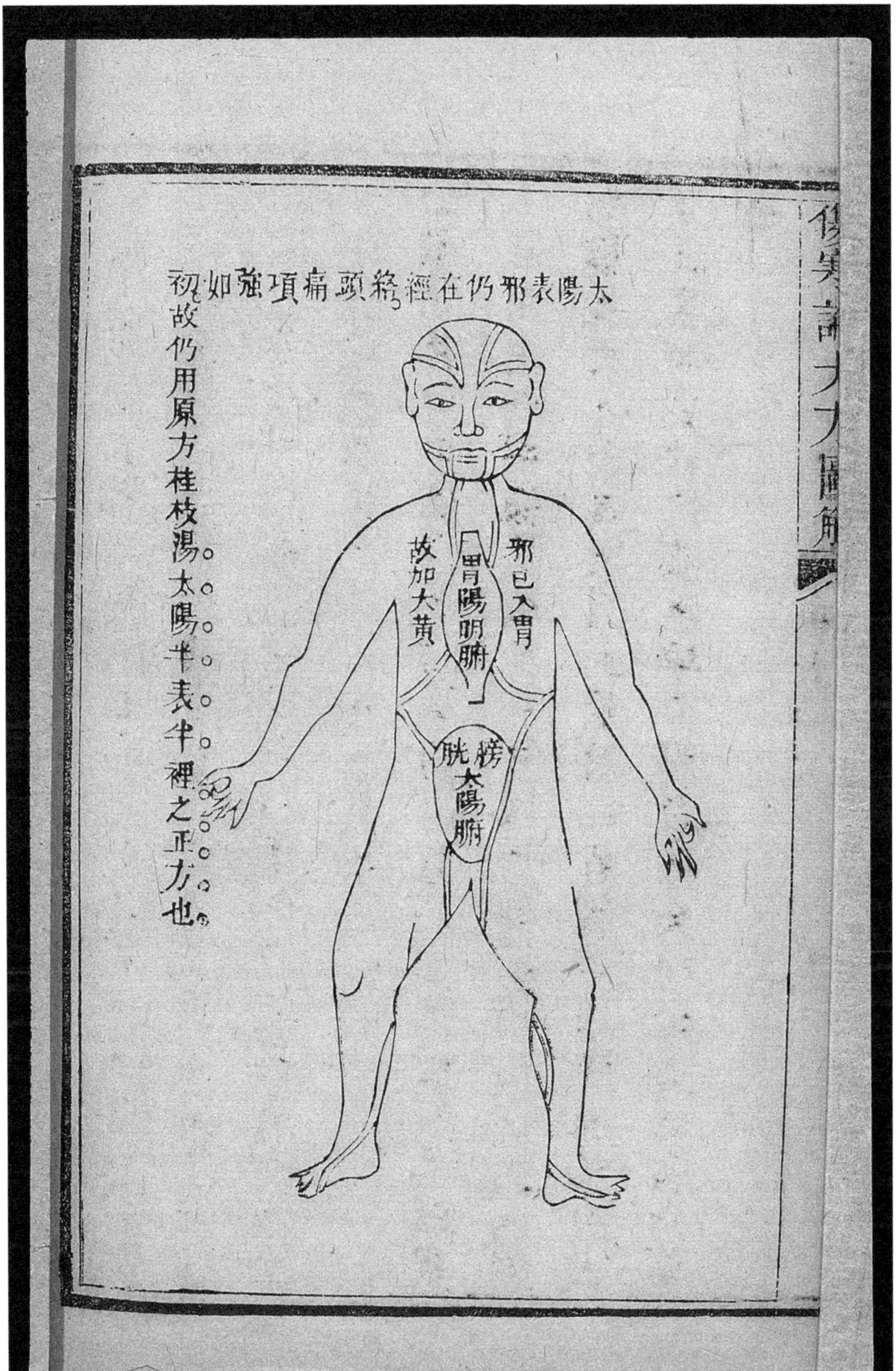

FIGURE 5.2 Woodcut illustration from an 1833 edition of the *Treatise on Cold Damage*. The image and text are about the diseases to be treated by rhubarb. Wellcome Images (https://wellcomecollection.org/works/amax2bs2). Attribution 4.0 International (CC BY 4.0).

arrived in Moscow, or any other European city, by land or sea, the physical nutmeg would be used in medicines according to the information about the textual nutmeg found in any number of Enlightenment works. Pierre Pomet's (1658–99) 1694 text *l'Histoire générale des drogues, traitant des plantes, des animaux et des minéraux, etc.* (translated into English in 1748 as *A Complete History of Drugs*), devoted two pages to nutmeg, as well as making repeated references to it throughout the text. Alongside a description of the medicinal properties of nutmeg—among a dizzying range of attributes, nutmeg seeds also "fortify the brain"—Pomet went into the geography of nutmeg's cultivation in Southeast Asia, and its route to Europe (Pomet, Lémery, and Pitton 1748: 128–30). The more intensive trading connections that held Afro-Eurasia together in the Early Modern period meant that the long textual history of nutmeg could be better mobilized by Europeans in their medical practices.

The plants of European Enlightenment medicine were not European. Or rather, they were not exclusively European, and even when they could be European, they were not unambiguously so. Building on the essential botanical contiguity of Europe with surrounding regions, and centuries-long trade contacts that stretched across Afro-Eurasia, Enlightenment Europeans were at least as likely to consume a plant that had grown on the Manchurian coast as on the Mediterranean.

NEW WORLD, NEW DRUGS

As post-1492 Eurasians came to terms with the new shape of the Early Modern world, they also began to come to terms with the new possibilities for medicinal plant remedies this presented, as with sassafras, the medicament from a Floridian tree that Tsar Mikhail Fedorovich was prescribed in Moscow in 1645. The excitement some Europeans had felt around a century earlier at the "New World" is expressed in one sixteenth-century English edition of a Spanish text on American medicinal plants: it styled itself as "Joyful News out of the New-Found World" (Frampton 1577). These new possibilities were primarily centered on the Americas. Despite the Dutch foray to Australia in 1606, Australian commodities, like eucalyptus (*Eucalyptus globulus*), only became known in Eurasia after the British-sponsored Cook voyage of 1770, and began to be exploited for European medicines and other practices from the nineteenth century on (Clarke 2008: 38). In contrast, from the middle of the sixteenth century, various American medicinal plants were being written about in Europe, notably in the works of the Spanish physician Nicolás Bautista Monardes (1493–1588), which were translated and reprinted across Early Modern Europe. One other notable, non-American, plant that was new to Eurasians came into the spotlight in this period: coffee. Coffee, the roasted bean taken from a berry-producing bush (*Coffea arabica*) native to the Ethiopian highlands, would become most famous as a recreational beverage, but early on its potential medicinal benefits were also discussed. Across the seventeenth and eighteenth centuries, new American drugs, and this one new North African drug, would be prepared, consumed, and written about by Eurasians.

Following 1492, various European peoples—notably the French, Spanish, Portuguese, British, and Dutch—traveled to what became known as the Americas, journeys motivated by the potential financial gains to be made from the commodities native to these new (for Early Modern Europeans) territories. Most famous of these commodities was South American silver. Chinese demand for silver, and Spanish attempts to feed that demand, most dramatically with the 1519 voyage by the Portuguese explorer Ferdinand Magellan

(1480–1521) that aimed to find a route from western Europe via the Spanish outposts in the Americas to East Asia, and the 1571 establishment of the port of Manilla to facilitate that trade, became a major fuel to the Early Modern global economy (Frank 1998). There were unintended consequences to this global circulation of treasure ships: the mid-seventeenth to the early eighteenth century has also been called the "Golden Age of Piracy" (Sherry 2009). As recent scholarship has only now begun to fully explore, alongside the importance of silver and other treasures, American medicinal plants were also a major part of European exploitation of the Americas. When European nations sent administrators to their new American possessions, those people needed medicines. Sourcing familiar European remedies was expensive and logistically problematic, and so they increasingly turned to Native American medicines. On their return to Europe, they spread news of these remedies, which information was recorded in medical books. Notable examples include the Monardes texts that aimed to popularize such remedies, in substantial part for the financial gain of the Spanish empire, which held de facto monopolies on many of these substances. Alongside Monardes' work, which was composed in a European language, some Euro-American texts of this period were originally written in American languages. Such works include the *Libellus de Medicinalibus Indorum Herbis* by the sixteenth-century author Martín de la Cruz (1552), a text about medical herbs, and *La Historia Universal de las Cosas de Nueva España*, also known as the *Florentine Codex* (composed between 1545 and 1590), a more general work on the history, religion, and botany of the region composed by various people under the direction of the Spanish Franciscan friar Bernardino de Sahagún (*c.* 1499–1590); both works were originally written in the Aztec language of Nahuatl, before being later translated into European languages. As Kapil Raj (2007) has argued for the South Asian context, what can seem from the finished texts to be western European scientific knowledge was in reality co-constructed by multiple actors across the Early Modern global world. Such co-constructed texts from the colonial American context helped Europeans appropriate American medicinal plants and local medical knowledge (Boumediene 2016).

This Atlantic world connection of western Afro-Eurasia and the east coast of the Americas that allowed Europeans access to American medicinal plants supplied a long-term demand for American plants. Trade records, prescriptions, and medical books, which circulated throughout the major centers of western Europe, list various American plants and their derivatives—sassafras from Florida; sarsaparilla from climbers (*Smilax* spp.) native from Brazil to southern North America; guaiacum from trees (*Guaiacum* spp.) native in tropical America; cinchona, also referred to during this period as Peruvian Bark or Jesuits' Bark and the source of modern quinine; and jalap from *Ipomoea purga*, a climber from Mexico—that testify to their use in Enlightenment medicine (Cook and Walker 2013; Finucci 2008; Griffin 2017; Huguet-Termes 2001; Klein and Pieters 2016; Wallis 2012). Such plants and their derivatives also appeared elsewhere in the Atlantic world, as for example at the West African port of Luanda in Angola (Gänger 2015). This interest was long lived: Patrick Wallis' work on the medical drug trade through London shows a substantial trade in American drugs from at least the late sixteenth century through to the late eighteenth century (Wallis 2012).

During the Enlightenment, American drugs also became known and used outside the Atlantic World. The presence of sassafras in prescriptions for Mikhail Fedorovich by 1645 speaks to a broader dissemination for American plants than their gateway into Eurasia via the westernmost part of Europe by the mid-seventeenth century. Indeed, alongside sassafras, the Russians imported a range of other American plants—notably

FIGURE 5.3 Cinchona, in Pierre Pomet's 1694 *l'Histoire générale des drogues*. Wellcome Images (https://wellcomecollection.org/works/usfg8m4r). Attribution 4.0 International (CC BY 4.0).

sarsaparilla, cinchona, and jalap—from the late seventeenth to the eighteenth century; they were also increasingly written about, in particular in eighteenth-century Russian medical works (Griffin 2017; Rowell 1978: 357). Recent work on medical drug traffic through the Baltic sea in the eighteenth and nineteenth centuries shows a substantial movement of sarsaparilla between western Europe and St. Petersburg (Veluwenkamp and Scheltjens 2017). By the late eighteenth century at least, American drugs were known and desired elsewhere in the Early Modern world: in the 1770s, the Sultan of Morocco used diplomatic contacts to source Spanish American medicaments (Gänger 2015). By 1791, American ginseng (*Panax quinquefolius*)—from the same genus as the Asian plants whose roots are so important to Chinese medicine—was readily available in East Asia, alongside local ones (Bian 2014: 196). A range of other American medical plants were also available around the Ottoman Empire, in the Japanese city of Nagasaki, and the Asian ports of Manila and Macau in southern China (Gänger 2015). Among the many medicaments from around the Early Modern global world that could be found in eighteenth-century Madras (present-day Chennai, in southeast India) was jalap (Chakrabarti 2005). By the late eighteenth century, plants which had only been known to Native Americans in the late fifteenth century were known to, and desired and used by, the trading world.

During the same period, as they were increasingly being exported to destinations across the Early Modern global world, American plants continued to be used in the Americas. An undated self-help book that circulated the Spanish-American colony of New Spain, *Receta especialisima contra calenturas, tercianas, dobles, o sencillas, aunque sean muy envejecidas* (A Very Special Recipe Against Fevers Tertian, Double, or Simple, Even If They Are Long-term) recommended, among other ingredients, well-ground quinine (Samayoa 2006: 8). In the sixteenth century, various Native American groups, including the Timucua of Florida, showed French and later Spanish invaders how to make an antifebrile sassafras root tea. The Rappahannocks of Virginia were similarly making sassafras root tea to treat fevers in the 1970s, and present-day Native American communities also use a sassafras root tea for recreational and medicinal purposes (Griffin 2020; Vogel 2013: 175). American plants also played a vital role in the healing practices of the unwilling migrants of the Early Modern Atlantic world: enslaved Africans. The Early Modern trade in plants and their derivatives was part of a broader system of commodity exchange, which included a trade in enslaved humans. Enslaved peoples, too, needed medicines. As shown by Tinde van Andel, the Maroon population of Dutch Suriname in South America, descendants of Africans who had escaped slavery, today use local American plants analogous to the medicinal plants used in their ancestors' place of origin in West Africa. Enslaved Africans conducted their own investigations of local American medicinal flora, recreating what they had known in their homelands (van Andel 2015). The new drugs of the New World were a long-term part of healing practices in the Americas—for both Native Americans and more recent immigrants—at the same time as having a substantial, and lasting, effect on Enlightenment Eurasian medical practice.

A new drug from the Old World also circulating the Early Modern global world was coffee. Apparently the seeds were first chewed, coffee only being consumed as a drink of roasted beans sometime in the fifteenth century, then known in the Ottoman Lands, Persia, and Europe by the late sixteenth century. It was the seventeenth century that saw its appropriately frenetic trajectory in Europe from curio in the early century to common beverage, and reason for the founding of the numerous and politically important coffee-houses, in the latter part of the century (see Chapter 2 of this volume), the first in Europe being in Oxford (1650). Mirroring and fueling that trajectory was a slightly less frenetic, but still rather busy, production of works appearing in manuscript and print in virtually every European language, and certainly in every European country, on the virtues and vices of coffee and coffee-drinkers (Ellis 2011: 12–24; Spary 2012: 51–95). In 1664, the British Physician Samuel Collins produced a text on coffee for the Moscow court, detailing the medical benefits of the drink including, of course, "*mnogosonie otgoniaet*"—"[it] drives away much sleep" (Collins 1664). Europeans then attempted to control the production of this desirable commodity, in somewhat the same way that they protected their monopoly on their colonial plants. Most famously, in the late sixteenth century the Dutch successfully transplanted coffee plants from the Horn of Africa to the Spice Islands, creating Java coffee. This new, Old World drug can be found in a variety of medical texts from Enlightenment Europe, but it was always more popular as a recreational commodity than as a medicinal one.

The denizens of the Enlightenment world had access to an increasingly wide range of medicinal plants. Many chose to take advantage of this. Others, conversely, took exception to the use of these new and foreign drugs. European invectives against both the American drugs, and coffee, were widespread, if apparently mostly ineffective in stemming the tide of these commodities' consumption. As discussed by Alix Cooper (2007), such an attitude often went hand in hand with a desire to return to Europe's natural wealth, to turn the renewed interest in medicinal botany back to the homeland, and exploit not foreign, but indigenous, nature. Similarly, during the eighteenth and nineteenth centuries, a number of Tokogawa physicians increasingly sought to find native Japanese replacements for imported medicines, and more broadly to develop native Japanese healing practices as an alternative to the Chinese medicine that had previously dominated Japanese medical practices (Trambaiolo 2013). In British India in the 1790s, there was a drive to commodify a local antifebrile as an alternative to importing American cinchona (Chakrabarti 2010). Although such attempts ultimately failed to remove completely foreign elements from the Enlightenment pharmacopoeia, they did promote the use of native plants to the extent that they, too, are found in trade records, prescriptions, and medical books throughout this period. Indeed, although the presence of American drugs is significant, the majority of plants in the Eurasian pharmacopoeia remained distinctly Old World well into the nineteenth century.

As the initial invasion of the Americas in 1492 was translated into a more permanent occupation across the Early Modern period, there was a massive exchange of medicinal plants from the new European colonies and outposts to Europe and beyond. This was significant. It shaped colonial efforts, and reshaped both elite and humble medical practices for centuries. Coffee, a new drug from the Old World, became a popular recreational beverage rather than a major medicinal commodity. This influx of the new and foreign prompted a backlash in certain quarters, with much effort being devoted to promoting indigenous nature as a source of medicinal (and other) bounty. The role of the globalizing Early Modern world in Enlightenment medicine was then to promote countervailing tendencies, and overall increase the breadth of medicinal plants in use.

MAKING DRUGS

In 1645, American sassafras, North African senna, Chinese rhubarb, and local juniper berries given to Tsar Mikhail Fedorovich were all processed into medicines: he was to take the various plants chopped up, as a mixture combined with Rhenish wine (imported from the German lands) twice a day for two days, and on the third day to drink the same herbs, but in heated mead, a locally produced honey-based alcoholic beverage (Mamonov 1881: 120). This combining of materials was a vital stage in the process of transforming plants from natural objects into medical drugs. The Early Modern period in particular developed a range of increasingly involved pharmaceutical processes used in creating medicines, processes which made a substantial impact upon medicinal plants as objects.

Cutting, grinding, heating, combining, distilling, extracting, drying, compounding. A fresh plant could go through many stages to be made into a medicine. At the very least it would be cooked in some ways, as in this recipe for making juniper spirit, written by the Russian Archbishop Afanasii of Kholmogory in 1696:

> Take juniper berries, pound them, and pour water or beer over them, heat the mixture, but not too hot, and put it in a wooden vessel, so that the mixture does not become too fluid, add hops [*Humulus lupulus*], and leave it to sour for two weeks. When it has soured, distill it and you will get arak, then add a few more berries, distill it again and you will have spirit.
>
> (*Reestr iz dokhturskikh nauk*, *Extract from Doctors' Knowledge*, 1696, quoted in Griffin 2015a)

Other processes were substantially more involved than the one Afanasii described: compound medicines involved tens of ingredients being combined. The most famous compound medicine was theriac, an ancient world medicine long popular in Early Modern Europe. In Italy, the creation of this medicine took months, and was a public ceremony. The importance assigned to the process was due to the particular nature of theriac as a compound medicine; it was made from more than forty ingredients, one of which—viper's flesh—was thought to be poisonous in its natural state. The processing of the ingredients had to take place correctly to ensure the medicine was safe, but also efficacious (Pugliano 2013). Theriac thus combined plant ingredients with a variety of other kinds of natural objects. Theriac's fate elsewhere in the Early Modern world was rather different: in China, it was originally lauded, but later condemned as a poison; in Russia, it was banned in the early seventeenth century, but enjoyed notable popularity by the early eighteenth (Griffin 2017; Nappi 2009a). Such compound medicines, created from many ingredients, remind us to contextualize the place of plant derivatives within medicine. As well as vipers' flesh, various other animal parts were commonly used in medicines worldwide, and chemicals were also used in various medicines. The use of chemicals was particularly significant from the seventeenth century onward, as the writings of the Swiss physician Paracelsus (1493–1541) promoting the use of chemicals in medicine were influential in both western Europe and the Ottoman Empire throughout the seventeenth and eighteenth centuries (Shefer-Mossensohn 2010: 177; Webster 2008). Plant ingredients were important to Enlightenment medicine, but they were far from the only category of natural object being used.

Correctly performing such complex techniques required education. This often came in the form of an apprenticeship to an established master, during which novices learnt much of their new craft through practical instruction. Most major western European cities had guilds of medical practitioners, in the same way that they had guilds of other artisans. The European creators of medicines were the apothecaries originally often combined in guilds with other purveyors, like grocers and candy makers, of consumable goods, but by the Enlightenment often having broken away; in London, in the late medieval period apothecaries were part of the Worshipful Company of Grocers (established 1345), only forming a separate guild, the Worshipful Society of Apothecaries, in 1617. These organizations helped police the boundaries of acceptable pharmaceutical practice. Other countries did not have the guild structure: in seventeenth- and eighteenth-century Russia, the Palace medical department, the

FIGURE 5.4 Nineteenth-century English medicine chest, including drugs such as rhubarb, jalap, and lavender. Wellcome Images (https://wellcomecollection.org/works/ysz9yw4d). Attribution 4.0 International (CC BY 4.0).

Apothecary Chancery, trained its own apothecaries from 1654 and licensed private practitioners from 1701; in the Spanish Empire, all medical education, including the apothecaries, was controlled by the royal *Protomedicato*, a kind of Early Modern Surgeon General (Griffin 2017; Samayoa 2006). Whether regulated by practitioners themselves, or by the increasingly ambitious Early Modern state, education provided an opportunity to regulate the practice of the drug-makers.

Practical pharmaceutical techniques were also the subject of craft texts. The earliest such European work was Hieronymus Brunschwig's (*c.* 1450–*c.* 1512) *Liber de arte distillandi* (Book on the Art of Distilling), first published in 1500 and then republished, copied, and translated across Europe, but Brunschwig was only the crest of a wave of European publications presenting textual and visual introductions to the techniques of the apothecary's art that proliferated across Early Modern and Enlightenment Europe. Similar works also appeared in print or in manuscript globally. One example would be the continuation of the East Asian *bencao* tradition, texts collecting knowledge on natural objects including medicaments. Li Shizhen's *Bencao Gangmu*, or *Compendium of Materia Medica*, was such a text, compiled in the late sixteenth century, and reworked and reprinted long after (Marcon 2015; Nappi 2009b). It was from texts like these, as well as from masters, that drug-makers learnt to turn plants into medicines.

Official medicine has left us many sources, and so is often the center of a medical historian's attention. It is important to remember, however, that this is an artifact of the distribution of literate activities, and the vagaries of document preservation, not an unproblematic representation of the world as it was. As noted by Andreas Renner, writing of eighteenth-century Russian medicine, official medicine is often treated as the norm, and unofficial medicine as the aberration. In reality, the opposite was closer to the truth. Unregulated, if not consciously criminal, healing practices were the near-unchallenged norm of medical practice for most people, most of the time. Official medicine, with its always insufficient numbers of practitioners and enforcers, could not hope to compete (Renner 2010: 214). Yet attempt to compete they did. Unofficial practitioners were faced with three major challenges. Firstly, pharmacopoeias. Cities produced these volumes, which specified which medicines were approved for sale, often with their ingredients and permitted prices also listed. Violation of these regulations was a serious matter. Secondly, either guild rules or state laws prohibited untrained and unlicensed apothecary practice. Thirdly, those laws were indeed enforced, with the unlucky practitioners who were caught being subject to the full force of the Early Modern state's legal and penitentiary processes. In Early Modern Russia, such unofficial practitioners were often caught up in witchcraft trials—Early Modern Russians saw herbs and plants as potential *materia magica* as much as potential *materia medica*—and the records of those trials provide some of the only details we have about those unofficial practices (Levin 2010). Like their official counterparts, this mass of unofficial practitioners also combined plants with a variety of other natural objects to make their medicines, but often the nature of their illicit practice makes the exact details of this obscure.

Tracking how plants were transformed from ingredients into medicines reveals important aspects of their role in Enlightenment medicine. Judging from extant recipes, only a proportion of medicines that contained plants were exclusively plant medicines. A great number and variety of medicines, official and unofficial, combined plants with parts of animals, human corpses, chemicals, and minerals to make compound remedies.

SELF-MEDICATION

In 1645, Tsar Mikhail Fedorovich had his palace physicians treat him; in the 1690s, his grandson, Tsar Peter the Great (1672–1725), was presented with a self-help pharmacy text (Griffin 2015b). Works aimed at laypeople—as opposed to practitioners—were a major genre of medical text, in particular during the Enlightenment. Two developments fueled this. Previously, literacy had been primarily restricted to the professionally literate, from bureaucrats in China's mandarin system to England's scriveners, or the culturally literate, like western Europe and East Asia's nobility, expected to be versed in the literary products of their culture as a part of their special social status. During the Early Modern period, an increasingly diverse segment of the worldwide population became literate; one of the genres of text they chose to read was medical works, in particular self-help pharmacy texts aimed at laypeople wishing to make and use medical drugs themselves. The second development was the long-term rise of printing in western Eurasia. China was printing medical texts in the eleventh century; Europe developed an analogous technology only in 1450, with the production of the Gutenburg Bible (Goldschmidt 2008). These technologies allowed for the creation of large numbers of the same work to be created more rapidly than by manuscript copying, and so facilitated the spread of popular works. Much of western Europe enthusiastically embraced the press, but both the Russian and Ottoman empires were more circumspect about this new technology, only putting it to substantial use from the early eighteenth century onward (Coşgel, Miceli, and Rubin 2012; Franklin 2010). As print took greater hold in western Eurasia in the Enlightenment, it helped provide lay readers with self-help medical texts.

In Enlightenment Europe, many such texts explicitly pointed out their lay purposes, claiming that they could be read and used by any person. British books commonly used the phrase "for [use by] the meanest capacity," meaning the poorest people in society; Russian works stated that they can be read "by every person," in both cases emphasizing that these are works for laypeople, rather than trained medical practitioners (Griffin 2015b: 725–6). Moreover, these were British works published in English, not Latin, to increase their readership beyond of the elite. Similarly, in 1729, the Japanese physicians Hayashi Ryōteki and Niwa Seihaku published a popular medical manual (*Fukyū ruihō*) based on Chinese sources but published in *kana* script—a syllabic writing system that represents Japanese in a simpler form than *kanji*, which is based on Chinese characters—to enhance its accessibility (Trambiolo 2013: 307). All these books were written under the assumption that the material they contained would be of interest to basically literate lay readers, not only the educated elite, or professional medical practitioners.

Looking at such self-help books, they list a variety of medicinal ingredients. There are chemicals and plants, multi-ingredient compound drugs and single-ingredient remedies known as simples. Often, the kind of natural object recommended by a text is related to the audience. The pharmacy text written for Tsar Peter the Great begins with a chapter on the medicinal properties of gold, followed by a chapter on silver; a different edition of the same text, with a humbler audience in mind, does not include these chapters (Griffin 2015b). Tracing the American plants in Eurasian self-help texts is instructive. These plants were necessarily expensive and tricky to source. Yet by at least the nineteenth century they were listed in self-help texts, as with the 1814 edition of *The English Physician Enlarged*, a reworking of Nicholas Culpeper's (1616–54) text *The English Physician* (1652), which recommends balsam of Peru (*Myroxylon balsamum*; Culpeper 1814: 382, 385). Plants, including expensive ones, were a part of lay medicinal texts during the Enlightenment.

Ingredients, medicines, texts, and practitioners, were all important parts of Enlightenment medicine. But thus far this chapter has mostly left aside the very reason for all this activity: the sufferer. Sufferers are fundamental to medical activities, but their perspective—the "patient's view," as Roy Porter dubbed it—is often hard to access, not least in early periods when non-professional writing and ego documents (documents written about and to oneself, like diaries) were not widespread (Porter 1985). Nevertheless, the attempt to find that view is important and can be productive. Extant diaries and letters from this period commonly deal with health and sickness. The Scottish mercenary Patrick Gordon (1635–99) of Auchleuchries, in service to the Polish-Lithuanian Commonwealth and the Russian Empire in the late seventeenth and early eighteenth centuries, mentioned the health and sickness of himself, his soldiers, and his family repeatedly, such as the prescription of c(h)amomile flowers (*Chamaemelum nobile*) for his daughter Catherine's headaches in 1689 (Gordon 2010: 268). A British woman wrote of her reproductive problems to her aunt in 1777, noting that her doctor was then giving her "much Medicine" (Lane 1985: 222). The correspondence of the Guatemalan physician Dr. Narciso Esparragosa y Gallardo (1759–1819) from 1797–1819 records various interactions between Esparragosa and his (often distant) patients; these letters show the patients themselves taking an active role in deciding what treatments they should receive (Samayoa 2006: 10–14). In the early nineteenth century, a bannerman from Beijing named Mu Qixian frequently recorded the importance of his mother in making decisions about the extended family's health and medical treatments (Bian 2014: 292–6). Indeed, such personal records provide in particular a vital glimpse of women's role in and reactions to medicinal drugs: regulated medicine globally was heavily dominated by, and often officially restricted to, male practitioners. As was the case in so many spheres of life in the Early Modern global world, women played a greater role in private, household, oral activities than in public, written activities. The opportunity to recover these otherwise lost voices make diaries and correspondence vital to the history of medicine.

In the increasingly literate environment of the Enlightenment, where more and more non-professional readers and writers emerged, texts aimed at this group provide a window into lay approaches to medicine. Healing oneself on the basis of a text written by an expert was a fairly popular activity across the Early Modern global world. Alongside works by practitioners aimed at laypeople, at least some of the words of at least some sufferers survive, bringing into greater focus groups that are under-represented by a study of medical texts alone, notably women. The medicinal ingredients described in such texts are varied. Yet plants, including expensive and exotic plants, were a major presence in those documents.

MEDICINE TRANSFORMED? NEW PLANTS, NEW MEDICINES

If history writing is about tracking and analyzing change versus continuity, Enlightenment medicine presents a particular challenge, a challenge well illustrated by the 1645 prescription to Tsar Mikhail Fedorovich with which this chapter began. In this document, as with many medical texts from the middle of the seventeenth through to the early nineteenth century, change looms large. Most striking is the importance of the development of Afro-Eurasian-American interactions across the post-1492 period. A great number of American drugs—like the sassafras given to Mikhail Fedorovich—came into Europe from the sixteenth century onward. From there they were traded on to Europe's contacts in Africa and Asia, and—most importantly—continued to be traded in those places throughout this period, playing major roles in medicine. The more intensive trade contacts between the two ends of Eurasia, facilitated by ongoing Central Asian land routes and the newer Littoral System, raised the importance and increased the usage of Asian commodities in European medicines. Technical developments led to a greater role of complex production processes in medicines, transforming plants into medicines. The rise of non-professional literacy fueled the creation of increasing numbers of medical texts aimed at a lay audience worldwide, as well as a proliferation of ego-documents somehow addressing the issues of plants and health. The view from 1815 was of a medicine full of new, foreign products encoded in new texts for lay audiences, who enthusiastically consumed them.

Yet continuities also abounded. Tsar Mikhail Fedorovich was prescribed one American plant, but several Old World plants, which had been known as medicaments across Afro-Eurasia for centuries. The prescription was originally composed in Latin, the traditional language of science and medicine in western Afro-Eurasia from Roman times. The death knell of Galenic humoralism never seems to have been truly tolled, and this form of therapy continued across the period. Enlightenment editions of ancient world texts like that of Dioscorides proliferated; and in East Asia, Chinese medicine and Ayurveda both continued to be practiced on the basis of recent reworkings of ancient canonical works. Compound remedies known since the ancient period, like theriac, maintained adherents across this period. American plants continued to be used by Native American medical practitioners. So, was medicine transformed during the Enlightenment? Not entirely. It was infused with new ingredients, new practices, and new ideas, but much of the essence of the thing remained true to earlier traditions.

CHAPTER SIX

Plants in Culture

STEPHEN BENDING

The symbolic and cultural use of plants in the eighteenth century is, of course, an impossibly large subject. This was a period which saw the emergence of increasingly sophisticated accounts of plant physiology and sexuality—themselves part of a larger Pan-European drive to understand the natural world in all its forms—and with it, the influx of "exotics" not only from North America, but from Asia and the Cape. By the end of the century, highly prized flowers and unusual plants were no longer confined to the botanical collections of an aristocratic elite, but had made their way into the gardens of the middling sort right across Europe. Alongside this commercial transformation of plant availability, however, a much longer-term set of cultural traditions which turn to plants as the occasion for moral wisdom and medicinal support should also be recognized.

In the Christian world the Bible provided endless similes which turn upon plants, and particularly flowers; in eighteenth-century China, Confucian models of self-cultivation might draw upon the flowers of the cherry tree or the fragrance of a scented orchid; while in both Hinduism and Buddhism the lotus is only the most striking example of a flower as the focus for religious understanding. And, while the Turkish "language of flowers" would join Ovid's metamorphic narcissus (*Narcissus*), hyacinth (*Hyacinthus*), and laurel (*Laurus nobilis*), as an easy talking point for European elites—we might set that alongside the continuing influence of plant folklore and a popular confidence in herbal remedies which remained powerfully in place throughout Europe and beyond. For all of the Enlightenment's taxonomic endeavors, that is, and for all of its urge to contain and explain, in the cultural imagination of the eighteenth century, plants resolutely resist stable and singular categorization. Certainly, in the work of both Linnaeus (1707–78) and of his popularizers—such as Erasmus Darwin (1731–1802)—there is a confidence in humankind's ability to observe, describe, and give order to the ever-expanding world of plants; but that sits alongside long-established and powerful imaginings of plants which work in quite different ways from taxonomic classification and the supposed singularity of scientific vision. Linnaeus might offer his eighteenth-century readers the language and intellectual structures of *Lilium*, *Poa*, and *Pyrus malus* (lily, grass, and apple [now *Malus domestica*]), but for those same readers Luke's biblical invitation to "consider the lilies, how they grow," Peter's assurance that "all flesh is as grass ... But the word of the Lord endureth forever," and Milton's evocation of forbidden fruit were likely to be amongst the most immediate and the most powerfully felt associations to come to mind. Away from letter-writing and "literary" texts this also holds true, with perhaps the majority of Early Modern texts on practical agriculture turning—repeatedly—to biblical quotation for their support (see, for example, Eliot 1760: 151–3, where, in a work otherwise studiously practical, he offered a long series of biblical quotations to justify the planting of trees, the creation of groves, etc; and for the practice more generally, see McRae 1996).

One reason to focus on plants, rather than other obviously material and symbolic objects, is because of the particular ways in which they are implicated in these widely held sets of ideas and assumptions about the purpose of the Creation, about a natural state, about man's relationship with the natural world. Repeatedly, and characteristically, individual plants are seen to be at once distinguished from and folded back into these conflicting and culturally constructed ideas of "nature." Turning from nature as a whole to plant life in particular—whether spiritually inflected as the Vegetable Creation or scientifically as Linnaeus' *Vegetabilia*—we can see how shifting taxonomies and imaginings are woven into the very ways in which people make sense of themselves and the world around them.

Certainly, eighteenth-century writers find it almost impossible not to think of buds, blossom, and decay in terms of their human correlatives, and to add the most prosaic example to this list—though still capable of announcing a heavy biblical baggage—one need only think of that ever-problematic category of the "weed"—amongst the most obvious examples of symbolic instability, of a category which marks out some plants as "natural" but as nevertheless unwelcome (for an example of this see Bailey 1875). Not simply in the wrong place, weeds are at once a part of "nature" and a category primarily articulating humankind's relationship with nature and understand of itself: what constitutes a weed remains a matter of context and debate (Crosby 2015: ch. 7). For many eighteenth-century writers the weed is as much a metaphor as a physicality, standing in with regularity for moral and spiritual failings which must be plucked out: in 1753 Lady Mary Wortley Montagu wrote that "vices and passions (which I am afraid are the natural product of the soil) demand perpetual weeding" (Wharncliffe 1837), while a decade later Mary Delany claimed that when the mind "turns to mere frivolous curiosity it will lead you astray, and instead of finding you are in the midst of roses and every desirable fruit and flower, you will be entangled amongst briars and nettles, and all sorts of noxious weeds" (Woolsey 1879).

The complications become apparent, however, when we return from the metaphorical to the physical world. For the unlettered, and notably for unlettered women, physical rather than symbolic weeds could also be central to life. No metaphor at all, the removal of weeds from the gardens of the rich might be all that kept a laborer from destitution (Bending 2013b: 17–18; Way 2006: 7). And yet, what drives that labor remains in some sense the metaphorical power of the weed: weeding was physical for those who labored in a garden, but for those who owned the garden, the need to weed was likely to be driven by ideas of order and by aesthetic ideals in which the weed—by definition—should have no place. And, just as the weed-infested field of the slovenly farmer lent itself to economic moralizing and spiritual invective throughout the Early Modern period, so physical weeds in a garden could act as a metaphor for one's sense of self and of how one would be perceived in public. In recognizing the physicality of engagements with the vegetable world, that is, we should recognize, too, the insistent urge to dematerialize the material object, to see the symbolic in the physical.

This shifting backwards and forwards between the physical and the metaphorical is characteristic to understandings of nature throughout the period; and indeed part of the problem is just how "nature" is to be understood. Marxist critic Raymond Williams' account of why the term is such a cultural problem remains valuable here, and drawing on his work the cultural historian Ludmilla Jordanova has rightly claimed that "Those who wrote about nature in the eighteenth and nineteenth centuries negotiated through their writings not only their relationships with their readers, but also their relationships with God, and with nature itself" (Jordanova 1986: 21; Williams 1976; see also Bowerbank 2004 and Thomas 1983). Is nature to be framed as God's creation; as those things against which man must struggle; as a harmonious system but one too complex for man to understand; as that part of the creation from which man separates itself or of which it understands itself to be a part? Is nature to be acknowledged as containing and demonstrating inherent moral goods, a hierarchical order in which humankind finds itself to be at the pinnacle, or a system of amoral relations to be mapped and defined in ever greater detail? As Jordanova goes on to claim, many of these questions arise in the form of oppositions and dichotomies, and while these may seem merely muddled and contradictory, they are important because "they drew on old, often classical notions ... were ... deeply embedded in ways of thinking [and] could be used without self-consciousness as habitual, customary ideas which structured patterns of thought" (Jordanova 1986: 86).

Deeply embedded in these accounts of nature, too, was the urge to analogize, to see the human in the non-human nature—the weed is of course a case in point. As Jordanova suggests, such analogies could seem so self-evident that they were hardly imagined to be such. But they were also a powerful means of marking humankind's difference from the world around it even as they appeared to acknowledge the similitude. Thus, for the late eighteenth-century dissenting minister and scientific experimenter, Joseph Priestley (1733–1804), analogy was a defining feature of humankind's rational distinction. Writing of both animate and inanimate nature, Priestley claimed: "it is of prodigious advantage in treating of inanimate things, or merely of brute animals, to introduce frequent allusions to human actions and sentiments, where any resemblance will make it natural. This converts everything we treat of into thinking and acting beings. We see life, sense, intelligence everywhere." As the cultural critic Maureen McNeil has argued, Priestley accepts here the centrality of analogy in attempts to make sense of the Creation, but this is no mere equivalence. For Priestley, what remains important is the consciousness of this act of understanding, a consciousness that this mode of intellectual vision is what marks out humankind from the world in which it finds itself (McNeil 1986: 198–9).

What muddies such confident claims for the primacy of the rational mind, however, is that the need to make them arises in large part in order to resist the lure of the bodily, the sensual, and worst of all, the base. If the animal creation seems easy to separate from rational man, the lure of sensual experience, of sight and of smell, have always the potential to drag humankind from high-minded thought to the mere pleasures of sense. In this, too, plants play a crucial role, with writers of all kinds often torn between, or carefully attempting to negotiate, an acute sense of plants as material objects and symbolic forms, as misguided pleasures in a fallen world and as the intimations of a world beyond the physical. The conflicting understandings and potentialities that these frames of reference imply might be merged or separated in different spaces and at different moments; and for that reason, in much of the writing of the eighteenth century, these kinds of instabilities become a primary focus of attention.

Such problems are not unique to the vegetable world, and as Priestley's example suggests, many writers would happily include both animal and vegetable nature in their analogies. Nevertheless, in much writing of the period, that urge to analogize seems far more insistent in relation to plants. In some sense this may simply speak of the sources used, and suggest that the literate classes' obsession with gentility makes associations with

Priestley's "brute creation" less comfortable. Certainly, an insistence on rationality as the marker that distinguishes man from the rest of nature means that, on the one hand, while there is a powerful urge to sympathize with animals (and by the end of the eighteenth century, especially with suffering animals) there is little willingness to associate oneself directly with the "brute creation" (Ellis 1996; Thomas 1983); on the other, analogies with the plant world can be comfortably framed as themselves a demonstration of the rationality that separates man from other animals: to think in such terms is to think rationally and spiritually. More tentatively, we might also argue that the widespread ability to shape, move, and control plants—alongside failures to do so—gives them a peculiar resonance for many who write of them and who engage with them physically. Thus, if the Enlightenment's engagement with plants is thought of only within narrowly defined accounts of empiricism, improvement, and taxonomy, that is to miss some of the period's larger debates about what it means to be human, how one thinks about society, and about how, as a sentient and apparently rational being, one must constantly negotiate between mind and body, between the physicalities of existence and their shifting symbolic potential.

While many of these concerns, of course, appear throughout the centuries, this chapter explores some of the inflections and anxieties that seem particularly pressing in the period. Valuable work has been undertaken in a wide range of academic disciplines on agriculture, on crop production and human productivity which might be framed as a georgic struggle against nature or a pastoral world of spontaneous natural production to serve man's end (notably Alpers 1996; Feingold 1978; Low 1985; McRae 1996; Thirsk 1985–9); and there has been a huge amount of work, too, on the use of planting in landscape gardens, on the language and practice of "improvement" in the shaping of the wider landscape on both sides of the Atlantic, as well as on the role of plants in the expansion of empire (Casid 2005; Chambers 1993; Crosby 2015; Laird 1999); but what interests me here is how the more public language of poetry and didactic literature, of biblical quotation and empirical study, might be negotiated not only in public but in private, not only by major writers and philosophical thinkers, but also by that larger group of literate individuals attempting to make sense of their lives. My focus, then, will be on some of the ways in which the empirical, the social, and the symbolic are negotiated in different forms of writing, and firstly to the potential of flowers to confront the individual with the relation between physical and metaphysical worlds, between public expectation and emotional reaction or spiritual response.

That flowers might seem amongst the more trivial of topics in an account of eighteenth-century plants already hints at how they can help the understanding of human evaluations of plant life (for a broad cultural history of which see Goody 1993). If the planting of trees and crops (the former, of course, often also a large crop, but also often imagined quite differently) tended throughout the eighteenth century to be associated with men, flowers, whether wild or enclosed within a garden, were more often associated with women. Delicate, intricate, seductive in beauty and scent, but ephemeral, gaudy, and slight—flowers were all too often an account of femininity and their description turned frequently on moral judgments and on the ways in which humans might value themselves.

Notoriously, the poet Jonathan Swift (1667–1745) ended "The Lady's Dressing Room" (1732) by likening women—or at least fashionable women and their make-up—to "Gaudy Tulips rais'd from Dung"; and while critical debate continues about whether the poem should be read as vicious misogyny or as the unpacking of a misogynist vision, the association between women and flowers, and the assumption that the association entailed a judgment, was powerfully present throughout the century. Arguably just as detrimental for women, flowers and flower gardens were drawn upon throughout the eighteenth century as a wholly conventional means of defining socially acceptable forms of femininity and of aligning it both with domesticity and piety. It is in this context that the Reverend John Bennett (fl. 1783–92) published his *Letters to a Young Lady* (1789), in which he notes first the fashion for botany as an "amusement" for the "elegant," claims that "Nothing is more calculated to amuse the mind, improve the health and spirits, and to inspire at once cheerfulness and devotion," but then quickly moves on to stress the value of these benefits for the policing of bourgeois femininity (Bennett 1789: 1:199). For Bennett, "Attention to a *garden* is truly [a] *feminine* amusement" (2:14), and, when safely confined to the study of shrubs and flowers, the idealized woman naturally emerges into bloom:

> ... she has never a moment *unemployed*. She is always *smiles*, because she is always innocent. Her pleasures are of the rational and *refined* kind. They never leave a thorn in the heart, or pluck one, blushing rose from her cheeks. How solid and how calm, if compared with the midnight revels of fashion, or the giddiness of admiration!Be like [this woman], my dear girl, and you will always be happy. Study nature, till it leads you up to nature's God. Pore on plants and flowers till they perfume you with *real* devotion; and I will engage you to become, in your turn, one of the most beautiful flowers in the creation.
>
> (1:200–1)

Much of this, indeed all of this, is utterly conventional, with the urge to naturalize women's affinity to flowers, but also with that tellingly swift movement from woman as polite scientific observer to woman as herself a flower (Shteir 1996). This is a transition rarely made in the same way about men, but it is powerfully present throughout the century, and it is one that we can see more fully if we turn to a work published in the middle of the eighteenth century, the Reverend James Hervey's *Reflections on a Flower-garden. In a Letter to a Lady* (1747), Hervey's opening gambit is, as his title suggests, to offer his readers an imaginary flower garden as the opportunity for a series of reflections on self, and it is a text worth exploring because it offers such a striking example of how a work insistently concerned with moral and spiritual readings of imaginary flowers can be so closely caught up in Enlightenment debates about how humankind should understand its place in the creation.

Hervey's popular but peculiar spiritual text does not simply provide a convenient counterpoint to the narratives of clear-sighted empiricism, of globalized movement of plants, of scientific endeavor, and of sustained horticultural and agricultural experimentation which tend to make up a conventional account of the Enlightenment's relation with the vegetable world. Rather, his insistently moral and religious work sets itself squarely within those larger debates about human nature, the natural world, and the relation between the two. Hervey (1714–58) provides us, that is, with some of the key dilemmas and points of negotiation for those who lived and sought to understand themselves in the modern world of the eighteenth century.

Central to Hervey's text is inevitably an analogic understanding of nature, and he took as his cue a passage from *The Spectator* VII, no. 477 in which Addison suggested that the gathering together of plants into a garden "gives us a great Insight into the Contrivances and Wisdom of Providence, and suggests innumerable Subjects for Meditation." Hervey's rather laborious attempt to delineate those subjects begins by offering his readers a flower garden set within a wider landscape. Implicit in this opening maneuver is the dividing of plants into—on the one hand—trees and crops (the latter framed by quotation from Virgil's *Georgics*, and linked not with labor here but with spontaneous growth) and—on the other—those flowers and shrubs which will become the focus of detailed attention. Standing on a terrace, Hervey's narrator looks out first to this larger landscape before turning to view the flowers within the garden walls. That movement, between the larger view and the detail finds its echo in descriptions of landscape throughout the century; it allows Hervey to offer another of the most quoted lines of poetry in the eighteenth century, Milton's "These are thy glorious Works, Parent of Good, / Almighty! Thine this universal Frame, / Thus wond'rous fair! Theyself how wond'rous then!" (*Paradise Lost*, Book V, ll.152–4); it establishes the importance of a Christian vision which Hervey elsewhere likens to the workings of an Evangelic telescope and microscope; but it also establishes the smaller scale of the flower garden, and detailed engagement with individual plants, as a moment of focused intensity—one which ultimately finds its echo in Blake's "Heaven in a wildflower."

What follows is again in many ways utterly conventional, with Hervey claiming that:

> To an attentive Mind the *Garden* turns *Preacher*, and its blooming Tenants are so many lively Sermons. What an engaging Pattern, and what an excellent Lesson, have we Here!—*So*, let the Redeemed of the LORD *look unto* JESUS, and be conformed to their Beloved. Let us all be *Heliotropes* (if I may use the Expression) to the Sun of Righteousness.
>
> (Hervey 1747: 91)

These conventions are worth establishing, however, not least because they highlight for us a powerful set of ideas and images with which writers far less insistently didactic than Hervey find they must grapple. Notably, the lily, the rose, and the tulip all make their appearance, and appear only in order to decay and die. The lily may be "the Queen of the gay Creation" with its "incomparable lustre" and an "Air of Dignity and Grandeur" but its "spotted Whiteness must quickly be tarnished, and the snowy Form defiled in the Dust"; to "a graceful Shape, and blooming Complexion, the *Rose* adds the most agreeable Perfume" so that people "are never weary of drinking in its Sweets" but soon it will "resign all those endearing Qualities; and hang neglected on its stem, or drop despised upon the Ground"; and the tulip with elegant and artful "profusion of Dyes" may lately have been the pride of the border but its radiant stripes will soon be "rudely blended, with common Mould" (101). The lesson of such mutability is clear:

> One could wish … these loveliest of the inanimate Race, a longer Existence: But in vain: They *fade*, almost as soon as they *flourish*: Within less than a Month their Glories are extinct. Let the Sun take a few more Journeys through the Sky; then visit this inchanting Walk, and you will find nothing but a wretched Wilderness of ragged or naked Stalks.—But O! (my Soul exults in the Thought) the *Garment* of celestial *Glory*, which shall ere-long array the reanimated Body, will never wax old …
>
> (94)

What is striking about this—insistently feminized—moralizing of flowers, however, is that for all the conventionality, these accounts imply detailed observation of specific flowers throughout the year. And while Hervey goes on to offer more and more conventionally moralized emblems (anger "like a prickly thorn," peevishness "like a stinging nettle," and so on), the urge to observe remains potently present. As Hervey notes:

> Not long ago, these curious Productions of the Spring were *coarse* and mis-shapen *Roots*. Had we opened the Earth, and beheld them in their Seed, how uncouth and contemptible had their Appearance been.—But now they are the *Boast* of Nature; the *Delight* of the Sons of Men; finished *Patterns* for Enamelling and Embroidery; outshining even the happiest Strokes of the Pencil …
>
> (96)

Inevitably Hervey continues by inviting his readers to see in this rebirth of plants an image "of the *Resurrection* of the *Just*, and the state of their reanimated Bodies!" (94), but if this analogic reading of plants would seem to imply a fixed position from which to understand one's place in the world, the effect is oddly different. The endlessly repeated cycles of bloom and decay hardly produce a settled or stable point of view. Instead, the repeated movement between states leaves Hervey's Christian reader in constant flux: the physical world is always to be seen through, and seen beyond, in ways that insist not just on the small-scale cycles of seasons but on the radical disjunction between humankind's understanding of its own time in relation to the infinity of God's universe. In this sense, while analogy would seem to offer a clear set of relations between humankind and the vegetable creation, the result is instead a radical switching backwards and forwards between different states: the physical is to be perceived only as a marker of the metaphysical and the metaphysical finds its clearest statements in the physical world. Like all gardens, then, Hervey's flower garden does more than simply bring together a group of plants: it announces human intervention and it emphasizes human vision, but it also insists that it is a space of intense expectation and acts as a repeated demonstration of the metaphysical instability that analogy implies.

If this formulation seems overly abstract or abstruse, it is worth recognizing that the problems represented here were a matter of everyday experience for eighteenth-century writers who drew on plant life in attempts to record their own spiritual life. Thus, for example, in the letters and diaries of the young Irish woman, Judith Ussher (*c.* 1790–8)—who died at the age of eighteeen—she wrote in the mid-1790s, that her mind has been

> greatly depressed and brought very low under a feeling sense of the great depravity of my nature, which seems engrafted therein, and, like a subtle serpent, to entwine round my inward parts. When walking a few days ago in a garden, a lily attracted my attention; then how did the thought sink deep into my soul, that the soul that appears with acceptance in the presence of the Most High, must be clad in garments as pure and white as that lovely flower. How did I, and do I, fear I shall never attain to the being clothed with these unspotted robes, which I am persuaded is the beauty of the renewed soul.
>
> (Usher 1871: July 27, 1796)

There is nothing of Hervey's uplifting rhetoric here, no sign of pleasure, and while Hervey claims that "One can scarce be melancholy, within the Atmosphere of Flowers" (164), Ussher tells us something quite different.

Like Hervey, Ussher in fact offers an example some way outside of the mainstream, with the spiritual insistence of both writers likely to be at odds with less strident forms of Anglican belief, even if the same preoccupations remained uppermost for many writers. Christopher Smart—himself thought unusual by many for his propensity to fall down in the street to pray—offers us in his poem, "On a Bed of Guernsey Lilies [*Nerine sarniensis*]" something close to Hervey when he saw in these flowers' late autumn bloom proof that for the "philosophic mind … / We never are deserted quite / 'Tis by succession of delight / That love supports his reign" (Smart 1764). But we can also turn to that most read of mid-eighteenth-century poets, William Cowper (1731–1800), for this same sense of delight along with a fear of abandonment. Thus, for example, in a passage from *The Task*—which led Jane Austen (1775–1817) to write of her new garden in Hampshire "I could not do without a syringa, for the sake of Cowper's line"—Cowper walks through a winter landscape imagining the flowers that will reappear in Spring (Austen 1997: 119). Describing first "Laburnum, rich / In streaming gold; syringa, ivory pure" along with "The jasmine, throwing wide here elegant sweets, / The deep dark green of whose unvarnish'd leaf / Makes more conspicuous, and illumines more / The bright profusion of her scatter'd stars," he then continues:

> These have been, and these shall be in their day
> And all this uniform, uncolour'd scene
> Shall be dismantled of its fleecy load,
> And flush into variety again.
> From dearth to plenty, and from death to life,
> Is Nature's progress, when she lectures man
> In heavenly truth; evincing, as she makes
> The grand transition, that there lives and works
> A soul in all things, and that soul is God.
> (From "The Winter's Walk at Noon," quoted in Cowper 1785: 239ff.)

This urge to see in plant life an image of hope is powerfully present in Christian writing, but so too is that sense of desperation and melancholic fear hinted at by Ussher; and for all of Cowper's pleasure in plant life it should be set alongside those images which also appear in his poetry—notably, for example, in "The Shrubbery"—where seeing the beauty of flowers simply confirms one's distance from God: "How ill the scene that offers rest, / And heart that cannot rest, agree!" (Cowper 1782: 344–5).

For all of these writers, however, the notion that plants provide humankind with lessons was axiomatic, and Hervey's engagement with that axiom sets his text firmly within Enlightenment debates about the purposes of nature. Indeed, one of the central and repeated questions posed by Hervey's work—though utterly rhetorical and with the answer in no apparent doubt—is, what is nature for? For Hervey, the most direct evidence that nature was made by God for man came not just in the form of vegetable creation, but more specifically in the beauty of flowers. Accordingly, Hervey argues first that, "Amongst all the Productions of the Third Creating-Day, this of Flowers seems to be peculiarly designed for Man: A Present, calculated in an especial Manner for his Use and Delight. Man has, as it were, the *Monopoly* of this Favour … " (Hervey 1747: 78), and he then continues:

> The *chief End* of these beautiful Appearances [i.e. flowers], Philosophers say, is to enfold and cherish the Embryo Seed; or to swathe the tender Body during its infant State.—But whatever is the chief End of Nature, 'tis certain, she never departs from the Design of *administering Delight to Mankind*. This is inseparably connected with her other Views.—Was it only to secure a reproductive Principle, what need of such *elegant Complications*? Why so much Art employed, and so many Decorations added?

> Why should such Vestments be prepared, richer than Brocades, more delicate than Lawns ...?—If the great Mother had no other Aim, than barely to accommodate her little Offspring, warm Flannel, or homely Fustian, would have served her Turn ...
>
> It seems plain then, that Flowers were endued with such inchanting *Graces*, for the *Pleasure* of *Man*; and, in pursuance of this original Intention, they still pay their Court to the human Race.
>
> (Hervey 1747: 79–80)

Flowers, for Hervey, only make sense when seen from a human perspective; and here we see in turn both the limits of Hervey's ability to digest current scientific thinking, and that seemingly unproblematic account of humankind as nature's pinnacle (Thomas 1983).

Hervey's insistently Christianizing vision of plant life—or at least his remorseless analogizing—was too much even for some eighteenth-century readers; but we can see that same normative imagining of flowers as a demonstration of God's beneficence to man and indeed that same urge to analogize in forms of writing with quite different agenda. Thus, for example, turning from Hervey's sustained analogizing to a work equally sustained by analogy, William Mason's four-book poem, *The English Garden*, one finds what was at the time a hugely popular poem equally concerned with reading plants as texts.

Appearing between 1772 and 1783, Mason's poem became one of the foremost statements on English landscape garden design in the second half of the century (Mason 1783). In Book IV of *The English Garden*, Mason (1724–97)—like Hervey—presents a flower garden, but instead of Hervey's direct Christian analogies or Cowper's hopes of a future life, Mason provides his readers with a politicized tale of female suffering turning on the events of the American War of Independence. Planting what should be a garden of love, the complex flower garden created by the love-sick hero heightens the effects of feminized sentimentalism and counterpoints the human suffering of war as two lovers, and two nations, are torn apart. In the wider landscape garden of Mason's poem, however, it is the plants themselves that are to articulate a political agenda: no longer merely a sensual backdrop, both the species and the placement of trees, signal the poem's political intent.

Dealing with wood, plants, and water, Mason's poem is centrally concerned with the shaping of plant life as a representation of both the natural and of the nation. Here, however, I want to focus on just one aspect of the poem, which is Mason's use of myth and symbol as he described the planting of trees in the "natural" (English) style. As Stephen Daniels has noted, trees always carry a heavy ideological freight, accruing mythic and symbolic associations from one century to the next. Amongst the most powerful were those surrounding the oak, and to see how that might be the case one only need think of that most patriotic of English songs, "Heart of Oak," or Alexander Pope's striking image of the oak trees of Windsor Forest falling into the Thames to become the bulwarks of Britain's naval empire. For Mason, however, there was also a more specific set of references for the oak which associated it with the Anglo-Saxon liberties made possible by the great forests of the ancient past; and he drew on these associations in one of the more strident uses of trees to explain and defend the power of an English constitutional system constructed—as Mason saw it—in the face of foreign tyranny and recent absolutist oppression (see Kliger 1952; Lovejoy 1932).

Linking oaks with aristocratic landowners, and avenues of trees with both the false taste and false politics of Louis XIV at Versailles, Mason describes the reshaping of an avenue into the groups and clumps of the English style as an image of political change.

Thus, at the centre of Book I is the description of an avenue of trees, and faced with its destruction Mason asks:

> ... where shall the Dryads fly
> That haunt yon antient Vista? Pity, sure,
> Will Spare the long cathedral isle of shade
> In which they sojourn; Taste were sacrilege,
> If, lifting there the axe, it dar'd invade
> Those spreading oaks that in fraternal files
> Have pair'd for centuries
>
> (I. 318–24)

Mason here drew on a long-established set of political associations that allowed Abraham Cowley to write, "Hail, old *Patrician* Trees, so great and good! / Hail, ye plebian undergrowth ..." ("Of Solitude"), but if, as so many eighteenth-century writers suggested, an avenue of trees resembles the nave of a cathedral and invites the same pious thoughts, in Mason's poem it also signals subjugation and tyrannical control, and must therefore be moved or felled:

> ... Cruel task,
> Yet needful. Trust me, tho' I bid thee strike,
> Reluctantly I bid thee: for my soul
> Holds dear an antient oak, nothing more dear;
> It is an antient friend.
>
> (I. 332–36)

Here, then, the poem offers us a sentimentalized emotional attachment to trees alongside an insistence on their iconic political significance (Baridon 1985: 89); and what unfolds is an extended and elaborate image of constitutional (rather than absolute) monarchy. Instead of appearing in unnatural rows, each oak finds its own place in the landscape, with "saplings tall, discreetly plac'd, / Before, between, behind, in scatter'd groups" breaking up the "obdurate line" of royal influence. Fewer trees may remain,

> Yet shall these few give to thy opening lawn
> That shadowy pomp, which only they can give:
> For parted now, in patriarchal pride,
> Each tree becomes the father of a tribe;
> And, o'er the stripling foliage, rising round,
> Towers with parental dignity supreme.
>
> (I. 342–7)

An image of political change in keeping with Mason's persuasions as a one-time republican, the spatial reorganization of trees provides an emblem of constitutional monarchy, of the reorganization of the state, though a reorganization based, nevertheless, on the age-old rights of the landowning elite.

Mason's heavily politicized reading of trees in the eighteenth century may seem oddly elaborate, but the willingness to associate trees—and the oak in particular—with forms of patriotic nationalism was widely held. It is that same set of associations that allowed his close friend, Horace Walpole (1717–97)—author of perhaps the most influential of all the eighteenth-century accounts of English gardens—to claim the private ownership of trees as the very basis of English garden design (Walpole 1780: 4:247–316). Where in

France, Walpole argued, the crown owned all trees above a certain size—and thus gave landowners the incentive only to chop them down when small—in England, landowners' property rights gave them a political independence which was the basis of liberty: the ability to plant trees for future generations, and the shaping of those trees into the "natural" style, was both possible because of, and an emblem of, that liberty.

Mason and Walpole's political rhetoric was open about its agenda, but equally highly charged political readings of plants were often far more subtle in their approach. Thus, for example, while Mason might offer the replanting of trees or the creation of a sentimentalized flower garden in order to play out his opposition to British policy in America, we can also see the powerful emotional and political role played by plants for those in America who experienced the turmoil of war.

In the last years of the American War, while Mason was completing *The English Garden*, Ann Eliza Bleecker (1752–83) was writing of a garden on the other side of the Atlantic. With the British army and its Native American allies threatening the Hudson river valley from the north, Bleecker, a member of a wealthy Dutch trading family in New York, wrote letters, tales, and poetry about the experience of abandoning and returning to Tomhanick, the family's country house in upstate New York (Bleecker 1793). It was in this context of threatened violence and destruction that "Return to Tomhanick" (1793: 260–2), a poem seemingly anodyne in its celebration of flowers, took on a peculiarly political charge. Like Cowper's "Winter Walk at Noon," this is a poem looking forward to the rebirth of the spring, and while it draws on a recognizable set of conventions from English poetry, its subject is temporally and geographically specific. A poem about home, about loss, and about return—standard subjects for any number of pastoral poets—the text turns heavily on the evocation of particular plants (Bleecker names twenty in a fifty-line poem, half of them flowers) and they become the focus for, and expression of, emotional attachment to place. However, what sounds like easy sentimentalism—"My *little garden* Flora hast thou kept, / And watch'd my *pinks* and *lilies* while I wept?"—becomes something far more tenuous not only because Bleecker wrote of home, having recently lost a young daughter in the war with Britain, but also because the old social structures which allowed a wealthy woman her pastoral ease might also be under threat in a newly established republic claiming democracy and equality for its white population (Giffin 1993). In this context, the poem's listing of flower after flower at Tomhanick becomes an uncertain celebration as the poem moves from its opening image of a garden in winter to this hoped-for return of familiar pleasures. From a garden "defac'd" by snow and the threat of violence, Bleecker offers herself hope with the unfolding of spring:

Yet soon fair *Spring* shall give another scene,
And yellow *cowslips* gild the level green;
My little orchard sprouting at each bow,
Fragrant with clust'ring blossoms deep shall glow
...
The rapid *humming bird*, with ruby breast,
Seeks the parterre with early *blue bells* drest,
...
Then shines the *lupin* bright with morning gems,
And sleep *poppies* nod upon their stems;
The humble *violet* and the dulcet *rose*,
The stately *lily* then, and *tulip* blows.

However, if this listing of plants, and this imaging of returning pleasures suggests the desire for restoration, for re-establishment, for confirmation that the world is the same, it also holds the fear of loss, a fear that this seasonal cycle will mark revolution not as repetition but as change. Nature's inevitable cycle means that all of these plants "shall" reappear, but as much of Bleecker's writing from this period acknowledged, people have changed: after the war, the return of easy pleasure may be no more than a pastoral fantasy (Giffin 1993).

What complicates still further this celebration of flowers not yet returned is that the very naming of flowers and garden plants—the honeysuckle and lupin, the poppy, violet, lily, tulip and rose—not only reinforces a vision of leisure, but marks also the failure to acknowledge other kinds of plants and planting. It is not simply that Bleecker celebrates flowers, rather, the very focus on flowers and a flower garden works to occlude all else around them, physically but also politically. When edible plants make their brief appearance, it is either—like the corn—to produce a pleasant sound, or to act—as with the gathering of watermelons—as a classic *sponta sua* image of laughing children who "bear the bulky fruit away." Not only does this deny a place for the laborers on whom Bleecker depended, but it highlights that much larger ideological vision of white settlers who failed to see in the planted landscape around them the agricultural practices of Native Americans framed only as savage and threatening.

Bleecker's celebration of garden plants is, then, a celebration of private property, of colonial expansion, of a wistful desire to maintain the social and economic status quo existing before the war. The narrow focus on individual plants excludes even as it celebrates; and the display of emotion employs a rhetoric of shared human responses to nature even as it relies upon, and reasserts, social and economic division. But what remains so striking about the poem is just how easily masked those political agenda might be, by the appeal to an intense emotional attachment to flowers, by an appeal to the natural as also the normative.

To achieve her effects, Bleecker puts huge weight on her lilies and roses, tulips and lupins, to do the work of emotional expression, to say what is felt. That rhetoric of shared human response, of intense emotion as at once sharply articulate and beyond articulation, is as powerfully present in Cowper's engagement with plants as it is in Bleecker's; but where Cowper wrote very much for the public, Bleecker's poems were originally written for a small social circle and the carefully shaped selection of poems for which she is now known appeared only after her death. It is this semi-public mode of writing—with specific readers in mind—that we can turn to next, because it produces its own kinds of problems particularly when writers adopt an apparently confessional mode, a mode often consciously articulating its failure to say the right thing. Cowper's letters once again provide numerous examples of this dilemma, but here I want to turn to a close contemporary of both Cowper and Bleecker, the poet and translator Elizabeth Carter (1717–1806).

Part of that circle of intellectual women which has become known as the Bluestockings, Carter's letters provide many of the plant tropes that we have already encountered in this chapter—from roses among nettles, to weeding out vices, to blossoming when cultivated—but here I want to focus on a brief exchange of letters with her close friend Elizabeth Montagu (1718–1800), an English socialite and intellectual, in which we find both the embrace of emotion and—far more openly than in Bleecker—a recognition that embrace must be finessed, or explained, or ironically undercut in order to be acceptable (perhaps even to herself). Writing to Montagu late in the summer of 1785 after a powerful storm had destroyed hedges and orchards, torn hops from their poles and lefts trees in the neighborhood “snapt in sunder,” Carter told her friend:

> I had the mortification lately to see two most beautiful elms, which used to shadow a stile on which I often repose, and read in my walks, irreparably demolished. I am so simple, that I grieve for the loss of every tree that shadowed my youth, as for the loss of a friend; but when people come to my age every day produces feelings of this kind; and nothing but the firm persuasion, that there is a better world, where neither sorrows, nor even uneasy sensations come, reconciles one to travel on with tolerable cheerfulness towards that bourn, to which all our steps tend.
>
> (Carter 1817: 3:253–4)

Carter, the translator of Epictetus, and one of the major intellectuals of her age, was anything but “simple,” and her correspondence shows her to have been an acute reader both of other people’s works and of herself. That need to frame herself as “simple” is therefore worth acknowledging, not least because it comes in the context of a long-term articulation of the theme. Six years earlier, she had also written of the wrench felt on the loss of trees from a landscape which was at once physical and emotional. Writing of the local farmers’ endless felling of trees—and thus implicitly setting herself apart from an agricultural engagement with plants—she recalls,

> I seldom walk out but I find a subject of lamentation in the fall of some favorite tree, beneath whose shade my youth has reposed and studied. This would be a very serious grievance to me, if I did not recollect how far advanced I am in my journey, I trust, to a country planted with trees of perpetual verdure, against which no feller shall ever level his destructive axe. At present, one must learn to be contented with that devastation and ruin, which is the sad lot of mortal existence.
>
> (Carter 1817: 3:106)

In both of these letters Carter offers the fallen tree as a moment which confronts one with a sense of oneself; and characteristically what she articulates is the recognition and the negotiation of alternative but coexisting states. In drawing on lost trees and a lost past Carter communicates the basis of her Christian faith, a kind of down-to-earth acceptance of one’s humanity, with all its failings, alongside an unshaken belief in the afterlife (O’Brien 2009: 56–65); but the constant need to ironize is its own announcement of the dilemma in which Carter finds herself, as, she implies, might any other Christian. Fallen trees might be a ready symbol for Christian faith in an afterlife, but, as Carter is acutely aware, her reaction might as easily be framed as a bodily, emotive engagement with the physicalities of the world—and for that reason it must be reshaped and reimagined even as it is owned. For Carter, plants become a key marker of this ambiguity, and if that epithet of “simple” suggests the desire for some kind of an unmediated, emotional engagement with nature, it was an engagement which Carter recognizes must also be schooled.

The confessional mode adopted with Montagu when Carter writes of lost trees is, then, a recognition of public norms and expected responses: terming herself “simple,” Carter suggests this to be both true and untrue, both embarrassing and to be embraced. In this sense, her letters highlight for us not only the constantly shifting perspectives which come into play in encounters with plants, but that the abstractly symbolic in no way precludes the sense of delight, and often wonder, for which they are also so often the occasion. However, the question of how to express or acknowledge that delight remains a central problem, and we see it rehearsed in even quite brief accounts of plants, such as the description Carter gave to Montagu of her small garden in Deal, Kent:

> My garden is absolutely *en friche*, but I hope before the end of the summer to see it blooming ... with roses and jessamines; and I have very magnificently ordered a wall to be built for their security ... I have a cherry-tree with half a dozen blossoms, and an apricot that never bears, and all this in a piece of ground at least as wide, and I think rather longer than your dressingroom.
>
> (Carter 1817: 1:187–8)

Writing in late spring, Carter's account of gathering together plants, both existing and potential, laughs at itself even as it acknowledges current pleasure and pleasures to come. The description of a small number of plants in a small space, this is also the description of intense space, signaled less, perhaps, by the expense of the surrounding wall, than by the ironic embracing of pleasure, by the need to laugh and self-deprecate. What might be framed as a *hortus conclusus*—or even as Hervey's meditative flower garden—becomes something rather different here. For, where both medieval garden and modern moralist suggest and expect pious meditation, Carter offers us instead ownership and expectation, worldly pleasures and human desire. Resisting and acknowledging that spiritualizing urge so clearly announced with her fallen trees, Carter offers us here instead the knowledge both that such pleasures are misguided and that the physicality and the ownership of plants brings intense delight.

Carter's short list of plants highlights another problem too. We have already seen how Bleecker's listing of plants might signal delight in the face of—and as the effacement of—anxiety; but Carter's briefest of lists both expresses and fails adequately to express her emotional attachment to plants. It highlighted their importance, but the list remains somehow inadequate, at once a marker of excitement, yet still only a list, and a short list at that. If this is Carter's problem, it might be seen as her readers', too, in that many of the records left to us of plants both in gardens and beyond them take just this form. That is, while it is tempting to think of lists and account books as impersonal and objective, that urge to list seen in both Bleecker and Carter, in Cowper and even in Hervey, suggests something altogether richer and more emotive (Fludernik 2016).

Certainly if we turn to the likes of Lady Jean Skipwith (1747/8–1826), at Prestwould, Virginia, planting huge numbers of trees, flowers, and shrubs in the years soon after Carter was writing of her own small garden in England, we are presented with extraordinarily detailed lists of planting and of plant gathering; and if those lists of plants do not speak quite as Carter or Bleecker speak, it is hard not to recognize in them forms of excitement and desire even as they fail to articulate such emotions in ways that can now be reconstructed (Martin 2001: 126–30). That task of reconstruction becomes much easier, however, if one turns from Skipwith to her near neighbour, Thomas Jefferson (1743–1826), not least because of Edwin Morris Betts' hugely valuable contextualizing of Jefferson's famous garden book (Betts 1944). Running from 1766 to 1824 the garden book plays neatly into that account of Jefferson as a man of the Enlightenment. Offering detailed tables and studious observations from day to day and month to month, the garden book would seem to align Jefferson with the empiricism of Enlightenment science, and with a public and political world of nation building which has no place for emotion. That is undoubtedly the rhetoric Jefferson himself adopted in many of his letters to friends and to political colleagues. But for Jefferson, too, plants hardly tell one story or are explained by a single vision, and the gardens at Monticello offer a useful example of how plants might take on quite different meanings in different contexts, even when those contexts are geographically close.

In the design of the Monticello's different parts, visitors were invited both to think differently and to see plants differently, from the long vegetable garden with its square beds and views out across Virginia, to the enclosed space, flower borders and kidney-shaped planting beds of the west lawn, to the seclusion of the woodland grove, shorn of low branches to provide shade and signaling a place of retreat. Recent work has characterized the vegetable planting at Monticello as a "Revolutionary" garden which set about the scientific pursuit of plants suitable for a new nation (Hatch 2012); but alongside that we might set the flower gardens of the west lawn with their more domestic agenda. Certainly, Jefferson's letters to the young women of the family suggest the continuation of his scientific and collecting interests in this area close to the house, but they also assume—just as can be seen in Bennett's *Letters to a Young Lady*—that flower gardens are a suitable domain for women, and botany a suitable science. In turn, in the years after Jefferson's death, these early childhood memories of the flowers at Monticello became a powerful part of the Jefferson myth, of the great statesman in retirement, concerned only with domestic affairs.

These different visions of plants and people extend beyond the formal designs for flowers, vegetable garden, and grove—and sometimes in ways not planned by Jefferson. Away from the mountaintop, recent archaeological work has found some evidence of the small-scale planting of Jefferson's slaves around their dwellings. Such archaeological evidence tells us little of the thoughts and feelings of those who planted what are likely to have been additional subsistence crops. But most of those enslaved men and women would have access to the same highly charged biblical language as their owner; might well have shared and exchanged seeds and plants (though on a far smaller scale than their master); and if they would be cut off from the kind of scientific discourse that gave Jefferson both purpose and delight, there is no reason to think that the growth of these plants did not provide pleasure as well as utility. That we do not know, and cannot now reconstruct emotional engagements with plants for this group of enslaved people tells us only of the kinds of records kept of their existence. What remains, those records of manual labor in the fields, of course provides us with a far more familiar story of hardship and exploitation, of crops as the object of attention, and of black men, women and children providing the bodies for the disembodied labor of estate records. Where black bodies do appear—as with the violence routinely recorded in the Monticello plantation books—that appearance tended to be sharply at odds with the kinds of scientific and pastoral visions at the top of the mountain.

In the case of Jefferson himself, things are of course quite different, with a wealth of material on which to draw; and while his scientific endeavors at Monticello still form a major part of the stories told about him, we can recognize, too, those same shifting perspectives that we have found elsewhere. Thus, for example, when Margaret Bayard Smith visited Jefferson in 1809, she recorded the following exchange:

> As we walked, he explained his future designs. "My long absence from this place, has left a wilderness around me;" "but you have returned" said I, "& the wildness shall blossom like the rose & you I hope will long sit beneath your own vine & your own fig-tree".
>
> (Jefferson 2004: Retirement Series 1: 386–401)

Bayard Smith's response to Jefferson brings us from scientific gardening back to the commonplaces of Christian reference with which we began: that Christian vision of wilderness transformed into garden, which is at once an account of physical terrain and a metaphor for the Christian soul. But Bayard Smith was attuned also, in that reference to the vine and the fig, to the language of retirement after national endeavor, of peace after war, of rest after labour—Micah 4:4: "but they shall all sit under their own vines and under their own fig trees, and no one shall make them afraid ..."—a quotation much favored too in the correspondence of George Washington.

In this context, Jefferson's repeated statements about life at Monticello—of the kind: "I am constantly in my garden ... as exclusively employed out of doors as I was within doors when at Washington"—while appearing to celebrate private life are also then highly politicized (Jefferson 2004: Retirement Series 1: 161–2). Nodding at the example of Cincinnatus, the Roman general, returning to his plow after saving the nation, for Jefferson (as indeed for Washington) planting and gardening become a public statement of private virtue, a demonstration that the true patriot readily gives up power for the sake of the republic. For all that, however, we should still be careful not simply to read the political into these statements of pleasure in planting: if expressions of that pleasure are carefully mediated as political and domestic virtue, that is something for which we have repeatedly seen the need in writings on both sides of the Atlantic. But the sense of pleasure, of delight in plants, still remains. It is this, quite as much as scientific or political endeavors, that ultimately made Jefferson a man of the Enlightenment, for the Enlightenment was never simply a matter of objectivity or empiricism absent of religion or free from emotion.

For writers throughout the century, however, the problem of how to articulate that emotion was always present and was particularly acute in relation to plants, whether large or small, cultivated or wild. Lists and labels are, of course, essential to the Linnaean categories with which we started and so too is the notion of objective observation; but as we have seen, a list is never merely a list, and a description is never simply a description, for both forms inevitably describe their author. Alongside that urge to list and to categorize, to observe and to describe, we have seen the self-ironizing of Elizabeth Carter, the spiritualizing of James Hervey, and the virtuous domestication of John Bennett, and in those maneuvers we have seen, too, the peculiar status of plants in the lives of eighteenth-century writers. We can end, however, with one final description of a flower, and with an author—Erasmus Darwin—whose two-part poem *The Botanic Garden* (1791) was once given as a gift by Jefferson to his son-in-law.

In 1784 Anna Seward recalled the following exchange with Darwin:

> Dr Darwin called here the other morning. We walked to Mr Saville's garden, accompanied by its owner. Talking about some rare and beautiful plants, Dr Darwin turned to me, and asked if I had seen the Calmia [*Kalmia*]. On my saying *no*, he continued—"it is a flower of such exquisite beauty, that would make you waste the summer's day in examining it:—you would forget the hour of dinner; all your senses would be absorbed in *one;* you would be all *eye*." I smiled, and asked him to describe it: "What, in the first place, was its colour?"—"Precisely that of a seraph's plume." We laughed, as he intended we should, at the *accuracy* of the description. He told us afterward, that he had heard much of the flower, but, as yet, had not seen it.
>
> (Seward 1811: 1: Letter IV)

FIGURE 6.1 Mary Vaux Walcott, *Mountain Laurel* (*Kalmia latifolia*), 1879. Watercolor on paper. Smithsonian American Art Museum, gift of the artist, 1970.355.559.

Taking its name from the Swedish naturalist Pehr Kalm (1716–79)—who sent samples to Linnaeus—*Kalmia latifolia*, or mountain laurel, is an evergreen shrub, native to North America. An exotic in England, for Jefferson it was part of a familiar planting palette. Seward's anecdote is so revealing, however, because it draws together both the empirical observation and the imaginative delight which was so much the experience of plants for so many eighteenth-century writers. Partly an account of the seductiveness of the exotic, the exchange is also centrally concerned with the nature of description; it offers transport, fantasy, and delight, even as it undercuts them by turning to humor, to the recognition that delight might be excessive, or foolish, or wrong. Darwin's description of a flower he has not seen offers us, that is, an image of scientific investigation, of literary and mythic frames of reference, of a delight that must be tempered by irony, and of an observer acutely aware of how they will be observed. As the anecdote also suggests, however, intellectual engagement may be the easy part for those who write of plants, with any number of literary and philosophical, scientific and religious frames of reference readily to hand; but what is also reached for is the thing that can not quite be said, the need, and the recognition of the need, for trees and flowers, grasses and shrubs, to carry the weight of how one feels as well as thinks about the world one inhabits.

CHAPTER SEVEN

Plants as Natural Ornaments

MARK LAIRD

The topic "Plantings" in the Age of Enlightenment has been covered elsewhere within a broad framework of Enlightenment Gardening that considers progress, change and continuities, improvement, fashion, and consumption (Symes 2013). Plantings in relationship to the expansion of technological and scientific knowledge have been further explored: on the one hand, as the development of the greenhouse, on the other, as the rise of the ideas of Carl Linnaeus (1707–78) (Laird 2014; Symes 2013: 86–7). Women's self-fashioning through gardening and planting offers yet another way of framing the flower garden, the collection of exotic plants, botanical art or collages, in terms of gender, emotion, accomplishment, and religious belief (Laird 2017a; Laird and Weisberg-Roberts 2009). Plants as Natural Ornaments is thus a complementary viewpoint that draws upon a history of the decorative arts—a history, focused on Europe, in which Enlightenment investigations/perceptions of nature were offset by age-old artifices.

EUROPEAN GARDENING WITH ORNAMENTAL PLANTS: 1650s–1720s

Aspects of the history of plants as natural ornaments in the Age of Enlightenment can be previewed as a potted history of parterre design at Cliveden on the River Thames in Berkshire, the estate of the Scottish nobleman, George Hamilton, 1st Earl of Orkney (1666–1737) (Laird 2013a: 318–22). Religion, national identity, climate, and economics were all potential determinants in how "ornamental vegetal motifs" were used throughout western Europe for two centuries. As just one example, hardline Protestantism and the eschewal of display (especially through flowers in worship) had led, from the time of the Interregnum, to an increasing preference in England for the plain grass "plat" (Goody 1993; Jacques 2017: 34–6). It would often come to replace the highly ornate *broderie* (embroidery parterre patterns in box, *Buxus sempervirens*) with *plate-bandes* (flower borders) in a French style.

At Cliveden, in the early 1720s, Lord Orkney abandoned any notion of a French-style parterre to set off the magnificence of William Winde's French-style architecture. Claude Desgots' proposals for Lord Orkney of 1713 had followed what his relative, André Le Nôtre (1613–1700), had perfected at Versailles as the dominant model of European gardening from the 1660s to 1700s (see Figure 7.1): *parterre*—"a flat unshaded expanse, square or rectangular, decorated with ornamental vegetal motifs" (Lamy and Olivesi 2013: 229). After another, rather simple, design for a grass parterre was rejected around 1720, Lord Orkney opted for the simplest solution of all—an unadorned lawn described thus: "I call it a quaker parter for it is very plain and yet I believe you will think it noble" (Laird 2013a: 320).

FIGURE 7.1 Étienne Allegrain, detail from *Promenade of Louis XIV in the Gardens of Versailles*, *c*. 1685. Oil on canvas, 234 × 296.5 cm, Châteaux de Versailles et de Trianon, MV 752. Photo by Fine Art Images/Heritage Images/Getty Images.

The emergence of the lawns of the "English landscape garden," or *le jardin anglais*, through designers such as Charles Bridgeman (1690–1738), working at Cliveden by 1723, followed one hundred years of "gestation" in intellectual, pictorial, and literary terms (Hunt and Willis 1988: 2). This distinctive national style was bound up with politics and nationhood, especially after wars with France and with the Acts of Union of 1707. Significantly, in 1740 one of Charles Bridgeman's amphitheaters at Cliveden was the setting for a first performance of Thomas Arne's masque *Alfred*, featuring the aria "Rule Britannia." All along, lawns, benefiting from the maritime English climate, had proved economic to maintain, and, as Horace Walpole later remarked, the smell of mown grass "green enough to disgust a Frenchman" had a kind of perennial patriotic redolence (Laird 2017b: 212).

A history of plants as natural ornaments during the Age of Enlightenment should not, however, fall into the habitual trap of following Walpole's divisive rhetoric. Flowers and flowering shrubs and trees (many from North America) assumed "Continental and Global" redolence in England over the long eighteenth century.

In *The History of the Modern Taste in Gardening* of 1780, Horace Walpole (1717–97) pitted the English "natural" style against the French, and all other forms of "formal," geometrical, or "regular" design. The credit for the English invention went to John Milton, William Kent and his followers, and to "Mr. Pope" (Alexander Pope, 1688–1744), whose "retiring and ... assembling" groves at Twickenham Walpole considered the inspiration for Kent's Carlton House, near St. James's Park (Walpole [1780] 1995: 47). Carlton House was home of the "patriot" Prince of Wales. Yet, strikingly, the only known plan in Alexander Pope's hand (see Figure 7.2), drawn in 1724 for Henrietta Howard at Marble Hill on the Thames at Richmond, points to nuances rather than binaries—the essence of planting design history. Although the poet's proposal was superseded by Charles Bridgeman's layout of 1724, Pope's sketch plan is key to understanding ornamental features of European planting. Pope's verbal designations on the plan owe much to both France and England: for example, "Parterres [with] Flow'rs" close to the villa in counterpoint to indigenous features such as a "Wilderness of Fruit trees." The disposition in groves of "Flowring Shrubs" (by the characteristically English "Mount" and

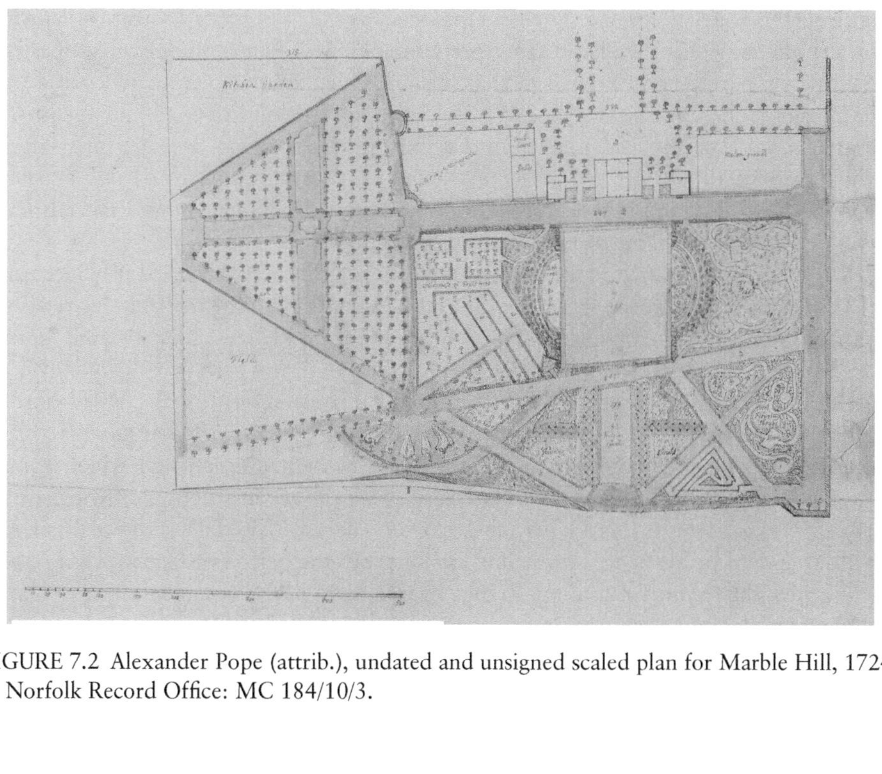

FIGURE 7.2 Alexander Pope (attrib.), undated and unsigned scaled plan for Marble Hill, 1724. © Norfolk Record Office: MC 184/10/3.

"Hanging Levell") anticipates the ornamental feature of the second half of the eighteenth century: the definitively English "shrubbery." (see Parker 2018: 133–5)

Apart from a degree of axial symmetry in the proposed layout—an axiality that survived in the Bridgeman scheme (and many later ornamental pleasure grounds in England)—plants as ornaments are featured in many "regular" ways: in hedgework or palisades, in arbors or *berceaux*, and in a maze that recalls the labyrinth traditions of Continental gardening up to and including Versailles (1666 onwards). A feature that is encompassed by a "yew hedge" looks rather like an interpretation of a French petal-shaped flower parterre of the style publicized by Jacques Boyceau or Claude Mollet. In short, geometries connect Pope to earlier gardening generations, notably that of John Evelyn (1620–1706) at Sayes Court with his radial parterre (Laird 1998: 181–7). And yet, as is made clear below, radial geometries were also reiterated in generations to come.

An earlier history of Marble Hill's setting along this stretch of the Thames would highlight the deep ancestral roots of the English style, going back to the time of Lucy Harrington, Countess of Bedford (1581–1627). In Twickenham, she had created a huge garden by 1609 that (according to a plan by Robert Smythson) was laid out in circles of birch and lime trees, with radial paths to mounts and to a maze-like armature of fruit trees, rosemary, topiary, and thorn hedge. These concentric rings seem to mirror a plan of the pre-Copernican universe (Strong 1979: 120–2). Such emblematic plantings would only gradually give way to an Enlightenment iconography: in William Kent's architectural works for Queen Caroline in Charles Bridgeman's Richmond Gardens along the Thames; and in Bridgeman's use of meadows at Richmond, following Joseph Addison's plea to use the "*natural* embroidery of the meadows" in place of the artificial *broderie* still kept up at Hampton Court as Anglo-Dutch modes of ornamentation under William and Mary.

Between 1650 and 1700, an Italian or Palladian impress was as profound as that from France or the Dutch Republic. Pope commended, within Henrietta Howard's Palladian villa, a room modeled on Palladio's interpretation of a Roman atrium. If Evelyn's parterre at Sayes Court drew upon French models, his grove evoked the evergreen groves of the Villa Borghese in Rome—the cypress, laurel, pine, myrtle, and olive. As he put it, with evergreens "an English Garden, even in the midst of Winter, shall appeare little inferior to the Italian" (Laird 1998: 190). Italy provided a *Georgics* aesthetic too, based on sound husbandry that united puritan and royalist in and after the Interregnum. France offered *la grande manière* and *ferme ornée* (Jacques 2017: 306–15).

The two decades following the successful conclusion of the War of Spanish Succession in 1713 brought confidence in new tastes, and a new lifestyle of intermittent retreat—both from politics and commerce—in the form of Italian "villas." John Erskine, Earl of Mar (1675–1732) was thus looking for a spot for his "villa" where eventually Howard's Marble Hill would be constructed. A conjectural reconstruction of the flower borders planned for the Earl of Mar by his gardener in 1708–14 for the Alloa estate in Scotland allows us to imagine the nature of a "British" use of flowers as ornaments, which is likely representative of Pan-European styles (Stewart 2016; the reconstruction drawings of the Earl's flower borders in Laird's private archive). The Earl of Mar's exile to France in 1716 put an end to his British gardening exploits, but the period of George I (1714–27) was to prove the testing-ground for a range of new ideas in the ornamental use of plants, including plants in tubs and pots in conservatories and in theatrical arrangement, whether indoors or outdoors.

GREENHOUSE TECHNOLOGY AND THEATRICAL DISPLAY: 1680s–1720s

The application of technology in the garden is nowhere as evident as in the cultivation of tender temperate and tropical exotic plants. However, during the century in which the Countess of Bedford, the Earl of Mar, and Alexander Pope were trying to set in motion "progressive" gardening along the Thames at Twickenham, "technological progress" was not as simple as a "linear progression." The early history of greenhouse and stove development is one in which the horticultural or ecological requirements of plants from vastly different geographical zones and habitats were insufficiently understood. Hampton Court, along another reach of the Thames, down from Marble Hill, provides the most complete record of conservatories, greenhouses and stoves in the period of William and Mary (1689–1702). At Hampton Court, today's recreation of the Lower Orangery display garden and the reconstructed Privy Garden show, respectively, the various ways in which plants were displayed as natural ornaments in the artifice of the tub or pot (Woudstra 2014).

By 1705, Mary Somerset (née Capel), 1st Duchess of Beaufort (1630–1715), kept a collection of tender exotics in her "Matchless Stoves" that were equal to those at Hampton Court. Dispensing with the latest Enlightenment technologies of Hampton Court (1689 Stoves; 1701 Glass Case), she still achieved prodigious results with succulent and tropical plants housed in a single structure of limited fenestration. An application of her hands-on knowledge of plants—what James Petiver (*c.* 1665–1718) characterized as "nursing care"—points to a broader and sometimes baffling question of the nature of horticultural knowledge in the Age of Enlightenment (Laird 2014: 67; 2015).

Inside the various structures for overwintering plants, a stage was an essential device for storage and display. Plants within stoves were placed on three or four or more shelves, with those plants requiring the greatest warmth directly situated on the floor that derived its heat from the furnaces below. Outside in summertime, it was usual to display all the tubs and pots on the level ground in regimented rows or ranks. While the French had turned this stage or stand into an ornamental feature—the "Theater of Flowers" (both outdoors and indoors)—the English had long adopted "theater" as a term to describe both the greenhouse stage and the complete collection that it supported (Hyde 2005: 74–5, 94–5; Laird 1999: 205). Thus, for example, Evelyn in his *Elysium Britannicum* (1650s onwards) wrote of the gardener disposing pots on "benches and shelves *Theatricaly* [*sic*] placed in degrees one above another" (Laird 1999: 205).

A group of flowers known as florists' flowers or flowers of distinction are linked to these early "theaters of flowers." In the case of auriculas, collectors paid particular attention to the color combinations found between the center or eye of the flower and the outer part of the petals. As Samuel Gilbert put it in *The Florists Vade-Mecum* of 1682:

> Their gold, their purples, scarlets, crimson dies,
> Their dark and lighter hair'd diversities.
> With all their pretty shades and Ornaments,
> Their parti-colour'd coats and pleasing scents.
> Gold laid on scarlet, silver on the blew
> With sparkling eyes to take the eyes of you.
>
> (Gilbert 1682: 54–5)

The "auricula theater" would emerge over the course of the eighteenth century as the quintessential way to display these flowers of distinction: tulips, hyacinths, and carnations, as well as auriculas. Commerce, competition, and pure caprice impelled practitioners and enthusiasts—those for whom the auricula was a livelihood or a hobby—to become dedicated to its culture. Such theatrical graduation would also emerge from the mid-eighteenth century as the organizing principle behind many a shrubbery or flowerbed in the English pleasure ground.

James Maddock (1718–86) described a particular "auricula theater" in his *Florist's Directory* of 1792. It was placed opposite a hyacinth bed with polyanthus pots. John Abercrombie (1726–1806) pointed out that such stages were not always in the form of Maddock's proscenium arch with a curtain; they could be "circular, having the shelves arranged in a pyramidal manner" (Laird 2000: 936). Here, in other words, was a concordance between the two types of "theatrical" graduation in the landscape garden: ranks of shrubs in straight rows (what we might call the school photograph mode) and concentric rings of flowers (what we might call the wedding cake mode). Equally striking is the correspondence to a form of interior display described by the London nurseryman Thomas Fairchild (1667–1729) in *The City Gardener* of 1722:

> The Chimneys which are generally dress'd in Summer with fading Bough Pots, might be as well adorned at once with living Plants, as I have observed at her Grace's the late excellent Dutchess of Beaufort.
>
> If one was to have a Pyramid of Shelves to be covered with Pots of blossoming Orange-Trees, with fruit upon them intermixt with Mirtles, Aloes, &c.
>
> (Blacker 2000: 60; Fairchild 1722; Hyde 2005: 92)

From the 1680s to 1720s, symmetrical flower arrangements prevailed: whether atop a *torchère* or in a basket; whether in a Delft urn or a Delftware fan-shaped vase; whether decorating a multi-tiered buffet in festoons or placed amid pyramids of fruit and sweetmeats on a table (see Figure 7.3); and whether in flower pagodas or tulip towers (Blacker 2000: 52–7). These would slowly give way to asymmetrical flower arrangements after 1740. In such free-flowing forms—an echo of flower paintings by Jan van Huysum (1682–1749)—some flowers were turned sideways and backwards, which lent a degree of "naturalness" to the artifice. A similar balance between naturalness and artifice is apparent in interior flower arrangements throughout the rest of the eighteenth century (Blacker 2000: 88–93). As the fashion for wider shelves to top a chimneypiece took hold, the room to display flowers in vases increased: a sequence of flower-filled vases—a garniture—could extend along a shelf in its symmetry from seven to eleven or even thirteen pieces. Alternatively, a hierarchy of baskets and vases of flowers could be constructed around a chimneypiece: from the fire grate to the chimney shelf to an elaborate mirror with bracket shelves for flower vases.

THEATRICAL GRADUATION AND FLOWER GARDEN GEOMETRIES: 1720s–1750s

In trying to appreciate the degree of artifice employed in planting the eighteenth-century flower garden—the preoccupation with neat gradation, the concern for order and diversity, the sensibility to multicolored effect, and the obsession with meticulous upkeep—it is not unhelpful to think of the contrivances of interior floral display or of

FIGURE 7.3 François Massialot, *Nouvelles Instructions pour les Confitures, les Liqueurs et les Fruits*, 1698, chapter entitled "Des Garnitures et Enjolivement des Services", 458. Bibliothèque nationale de France.

the contrived art of the "auricula theater." The theatrical shrubberies of the landscape garden, displaying individual exotics in six graduated ranks and in the seven hues of the rainbow, required almost as much finicky work as the "blooming stage." And just as a perfect show at the "auricula theater" depended on a "succession of proper plants in bloom," so too the ideal flower border required plenty of reserve pots for "plunging" (pots submerged in earth). Industry with knife, trowel and dibble, craft with stick, stake and twine, made the eighteenth-century flowerbed a work of art in a setting of nature (Laird 1999: 210).

In addition to graduation within a circular flowerbed (sometimes edged with sand, as seen in a painting of Grove House in the 1730s; Laird 1999: 185–203), there were other geometrical constructions to show off flowers. A radial petal-like form has already been mentioned in association with Alexander Pope at Marble Hill. So, too, Batty Langley promoted a radial "Flower Garden" in a grove in his *New Principles of Gardening* of 1728. Eight segments, like those of a cut orange, were disposed across the radius, with a circular bed at the center. Thomas Wright promoted "rosaries" that involved much more complicated construction lines. A single rose-garden of intricate geometric spaces (rather like a "rose window") might have been intended by the one sketch for Lord Barrington at Beckett Park of the 1750s. By comparison, his alternative sketch is more easily translated into "clumps" or "studs" on lawn. The fine construction lines—the interlocking circles—are quite distinct from the outlines of the beds. These lines, once removed, leave a layout of beds making up an entire garden, much as Wright would propose for St. James's Park in 1766. Thomas Wright's use of arc- or crescent-shaped beds in that 1766 plan appears to go back to his earlier sketches, in which the crescents are derived from Robert Furber's "Borders of Cut Work" of 1727. So, too, they anticipate the kidney-shaped beds Wright would devise for Netheravon in 1760 and that became a feature in the flower gardens of Nuneham Courtenay and Hartwell House in the later eighteenth century.

In looking at the arrangement of dishes on a dining table or buffet (see again Figure 7.3), we can see that the use of symmetry and geometry was similar to that of the flower garden. So equally, after 1750, the "natural ornamentation" of English pleasure grounds began to infiltrate the dessert table. Richard Bradley in his *Family Dictionary* of 1725 had given broad guidance: "the whole desert [*sic*] is to be set out with Flowers, Green and other Ornaments, according to the Season ... most especially in the void Spaces at Intervals" (Blacker 2000: 63). Yet Horace Walpole pointed out in 1753 that the previous "jellies, biscuits, sugar plums and creams" of dessert had "long since given way to Turks, Chinese and Shepherdesses of Saxon china ... wandering on the table, unconnected, among groves of curled paper and silk flowers" (Blacker 2000: 101). In 1756, another commentator, William Farrington, referred to the dessert as "a beautiful Park, round the Edge was a *Plantation of Flowering Shrubs*" (Blacker 2000: 101) This "plantation" alluded to new and fashionable "shrubberies," which, along with groves, were key features of most English pleasure grounds.

ORNAMENTAL PLANTINGS IN LANCELOT BROWN'S PLEASURE GROUNDS

Lancelot "Capability" Brown (1716–83) is known today as the most famous landscape designer in English history. He is associated primarily with large parks of expansive turf, with serpentine lakes, and with parkland "clumps" (circles of trees) and perimeter "belts" (long, thin or thick plantations). Yet most of his early commissions involved transforming the setting of the house, which was known as the pleasure ground. In the plan, contracts, and nursery bills (1752–6), for Petworth, Sussex, he demonstrated a skill in adapting geometric gardens to new geometries. For example, he was contracted to "finish the Parterre in front of the Green House" (Laird 1999: 133–41). The 1752 plan shows a grass parterre and greenhouse garden of bold hippodrome form that replaced the rectilinear parterre. A long rectangular "Aloe Garden" (likely for succulent display) and a near-oval "Bay Garden" (likely for evergreens including *Laurus nobilis*) perpetuated the traditions of exotic display inherited from generations before him. Only in a serpentine walk with "clumps" of shrubs can the full shift to irregularity be detected as "shrubbery." It was probably here that extensive supplies of exotic trees and shrubs from nurseryman John Williamson (fl. 1750s–1766) were employed, and with flower embellishments on the margins.

Brown's unimplemented design for the Duke and Duchess of Beaufort's pleasure ground at Badminton House, Gloucestershire, *c.* 1752, is equally suggestive of ways that landscape designers modified and adapted axial layouts (see Brown and Williamson 2016: 38, illus. 42). It may be compared to Thomas Wright's proposal of 1750 for the same garden area on the east front of Badminton as also to Wright's design for the Duchess of Beaufort at Netheravon, Wiltshire, of 1760 (Laird 1999: 128, 195). In 1750, Thomas Wright had proposed a complex assortment of planted compartments for Badminton, including an evergreen ellipse as a symmetrical cross-axis to the long axis from the east front. By comparison, Brown unified his composition around a broader "opening-and-closing" axis, weaving a sinuous circuit path through parterres and open

FIGURE 7.4 Mark Laird, reconstruction drawing of the effect intended by Joseph Spence's plan for Dean Paul's garden in Ireland, 1765.

groves. The parterres can be interpreted as simple grass "plats" enclosed by continuous flower borders. The groves were to be under-planted: on one side as a circle and on the other as an ellipse. These regular under-plantings were in the form of "tufts" beneath tree stems as also in "clumps." Perhaps it would have been similar to what appears in a painting of Grove House, Berkshire, dated to the 1730s (see above). A closer equivalent is suggested by my reconstruction drawing of Joseph Spence's proposal for Dean Paul's garden of 1765 (see Figure 7.4). By contrast, Thomas Wright's Netheravon proposal of 1760, while retaining a viewing axis (from temple or alcove seat to elliptical pool), shifted the balance in favor of a circuit around a central space studded with irregular flower clumps.

That Brown had adopted a "theatrical" approach to shrub and flower planting by 1752 might be assumed from his sketch proposal for a temple, placed on a viewing mount at Badminton. It appears to show a spiral-form ascending path (see Figure 7.5). The planting components in the sketch involve little groves of standard trees on lawn, with a regular dotting of flowers on that lawn, and with shrub clumps that are conical or "theatrical." Batty Langley had detailed a "cone of evergreens" in his 1728 grove plan, and perhaps Brown adapted that conical prototype to shrubbery clumps. Whatever his intention, the effect suggests a high degree of artifice.

Brown's plan of 1771 for Charles Pelham's Brocklesby, Lincolnshire, is valuable in showing another way in which regular geometries could be replaced by a different kind of geometry. The original geometric bastion gardens were shown by dotted lines, and, over the top of that trace, Brown sketched in a regular, octagonal flower garden in front

FIGURE 7.5 Mark Laird, an interpretation of the right-hand part of Lancelot Brown's unexecuted design for a temple, to be erected on an earlier viewing mount at Badminton House, Gloucestershire, *c.* 1752 (see Brown and Williamson 2016: 78–9, illus. 42 for the whole design and illus. 43 for the temple on the mount with planting).

of a new greenhouse. Petal-shaped flowerbeds, some up to 15 meters (50 ft) in length, were arranged around a central ellipse. The model for such florists' flower gardens—a specialized arena for the cultivation of flowers of distinction such at tulips, hyacinths, auriculas, and carnations—can be traced all the way back to the Renaissance. Richard Symonds' sketch of Pierre Morin's garden in Paris in 1649 shows a similar number of radiating petals to those in Brown's Brocklesby proposal (around twenty to twenty-four). William Chambers had created a more immediate prototype at Kew, near London, in 1762, and significantly the "petals" numbered twenty (Laird 1999: 202–3).

WILLIAM CHAMBERS TO HORACE WALPOLE: GARNISHINGS AND FESTOONS

Around 1700, when geometrical or "formal" gardens were still predominant in England, the writer Cassandra Willoughby (1670–1735, daughter of Francis Willoughby FRS and the Duchess of Chandos at Canons) described the 1st Duchess of Beaufort's Badminton wilderness thus: "Ye earth is covered with a variety of plants and primroses, periwinkle, etc." (Laird 2015: 189–90). Here was a way of augmenting the natural ground flora of a grove with artificial "embroidery" or "enamel": a dressing-up with cultivated flowers. At Wrest Park, Bedfordshire, Willoughby also noted: "to run up ye bodys of ye Trees are planted Honeysuckles and Sweet Brire." This dressing up of a tree trunk would survive the shift from regular to irregular gardening and acquired the label "garnishing." Batty Langley (1696–1751), who worked at Wrest Park, wrote in his *New Principles of Gardening* of 1728: "That all the Trees of your shady Walks and Groves be planted with Sweet-Brier, White Jessamine, and Honey-Suckles, environ'd at Bottom with a small Circle of Dwarf-Stock, Candy-Turf [*sic*] and Pinks."

Similar effects are evident in an anonymous painting of Richard Bateman's garden at Grove House, Old Windsor, which appears to date from the 1730s (see above). Mary Delany (1700–88) wrote of trees "embroidered with woodbine and the '*flaunting eglantine*'" (Laird 2015: 190–2). Paul Sandby's views of Horace Walpole's Strawberry Hill show similar effects, even through the changes of the 1760s. Hence, in 1765, Walpole could write of honeysuckles dangling in "festoons" from every tree. "Garnishing" was the term publicized by William Chambers (1723–96) to describe this gardening artifice. For example, in his manuscript in the Royal Academy library, Chambers highlighted everlasting pea, nasturtium, and sweet briar as all "fit for Garnishing Stems of Trees."

In architectural terms, a festoon is a carved, modeled, or painted garland of flowers, fruit, or leaves, suspended in a curve between two points. It differs from a swag, in which the object suspended resembles a cloth. In the early 1760s, in the Painted Room at Spencer House, London, James "Athenian" Stuart (1713–88) had *trompe l'oeil* garlands painted to adorn the frieze and to festoon the picture frames, as they would have done on festive occasions (Blacker 2000: 105–13). Garlands of real flowers, fruits, and leaves were traditionally made for celebrations throughout the year. A late-eighteenth-century version of a thick garland might have used unwaxed sash cord and mossing wire, with box foliage and spring flowers: anemones, tulips, auriculas, and guelder rose (*Viburnum opulus*). A summer version of a thin garland could entail using ribbon and mossing wire, with myrtle foliage, feverfew, and honeysuckles. Ivy and rose garlands might be entwined around classical columns to create a natural spiral that resembled the garnishing of tree stems. Thus "embroidering" came to represent different ways of dressing up or ornamenting the garden, which were derived from the decorative arts as much as from gardening.

Where *broderie* in a parterre had taken its inspiration from the silk embroidery of dress and waistcoat, "embroidery" in a grove employed nature's own artifice in imitation of architectural festoons and garlands.

GROWING INDOOR BULBS: THE CERAMICS OF SÈVRES AND WEDGWOOD

An instance of the decorative arts responding to a motif in the pleasure ground may be seen, perhaps, in the development of ceramic vessels for displaying flowering bulbs. Many were designed to hold double hyacinths fashionable in the mid- to late eighteenth century. From Robert Furber's pioneering experiments of the 1730s to John Abercrombie's writings of the 1780s, hyacinth bulbs were grown as singles in glasses and pots, or in larger ceramic containers with flat lids (Blacker 2000: 80–2; Hyde 2005: 93–5; Laird 2000: 935–7). George Voorhelm's *Traité sur la Jacinte* of 1773 illustrates singles in a water-filled carafe and an earth-filled Delftware pot. *Piedestal à l'Oignons* were bulb pots made by Sèvres from 1756 to 1773. The pot was filled with water and a single bulb placed on the detachable, gilt-edged tray that fitted into the top. Thereafter tiered vessels that held several bulbs began to appear in Strasbourg and Paris, just when the "theatrical" mode of pleasure ground planting was becoming fashionable on the Continent. A triple-tiered porcelain vessel, made at the Locré factory, Paris, *c.* 1772–80, allowed for a "theatrical display" of double hyacinths. The beautiful conical blooms, combined with the repeated rhythm of the supports and the sinuous leaves, would have established a pleasing Rococo ensemble.

It is misleading, however, to push the argument beyond the evidence. In England, where graduated flower clumps and graduated serpentine shrubberies were reaching their zenith from the 1770s to the 1790s, many modish vessels were not graduated. For example, Josiah Wedgwood simply adapted flower bricks as flat containers for bulbs, notably in caneware. He made them in different sizes with rounded cups in which the bulb might rest, and with small holes for supporting sticks. A bulb pot made at the Pinxton factory in Derbyshire, *c.* 1796–1813, for narcissus bulbs with supporting sticks represents a later stage of neo-classicism and a departure from the Rococo vessels of the mid-eighteenth century. Meanwhile, "naturalism" might have little to do with imitating the natural style and much more to do with a love of natural history or the "curiosity" of earlier generations. For example, the hedgehog crocus pot in the natural form of that animal, with holes between the bristles, looked back to ceramics of the past. Amusingly, it seems to anticipate the close-to-nature look of chia hedgehogs of recent times.

FRENCH *CORBEILLE*, ENGLISH FLOWER BASKETS, AND DUTCH STILL LIFE

In 1766, Lady Sophia Shelburne visited Blenheim, where she "walk'd about the Garden and saw the work of Mr Browne ... just before the windows is a pretty clump of flowers call'd a French basket" (quoted in Brown and Williamson 2016: 100). Since no representation of this clump of flowers survives, there is no way of knowing whether it corresponds to what the French called a *corbeille*. It is perhaps unlikely to connect to anything in the repertoire of the great Lancelot "Capability" Brown. On the other hand, it does correspond to the use of baskets of flowers in the interiors of country houses, where Brown landscaped or Chambers designed. For example, at Osterley Park House, in Isleworth just west of London, the long gallery runs the full length of the garden front (Blacker 2000: 90–2). Matthew Hillyard had decorated this gallery in the 1750s for Francis Child. On the marble chimneypieces, designed by Sir William Chambers, baskets of flowers (with metal liners and wire netting supports) would have been placed at either end, as though on the heads of female "herm" supports.

By the 1770s, illustrations of *corbeille* in French publications must have circulated in Britain, and these are likely to have inspired designers in the period just after Brown: for example, Georges Louis Le Rouge's *Détail de nouveaux jardins à la mode* (1776–88), which depicts the round *corbeille* in the duc de Biron's garden; or plate 369 in André Jacques Roubo's *L'Art du menuisier* (1775), which shows a multilevel *corbeille de terre* in plan and elevation (Laird 2000: 936–7). In the latter, the highly ornamental plant container, held up by a complex quatrefoil trellis or lattice work and spilling over with flowers, forms an architectural hierarchy, considerably more artificial than "theatrical" flowerbeds in the English landscape garden. A two-tiered flower stand designed about 1770 by Robert Adam for Osterley might be mentioned as yet another way of creating a "theatrical" effect—a natural artifice. Meanwhile, baskets for mignonettes (*Reseda odorata*) and other flowers were illustrated flanking a flower stand on the trade card of "basket makers and turners" in 1796, underlining the relationships among different forms of flower display (see Figure 7.6).

Humphry Repton (1752–1818), successor to Brown as the leading landscape consultant in late Georgian and Regency England, used the term *corbeille* in his Red Book for Courteenhall, Northamptonshire. Lady Wake's Courteenhall flower garden included, alongside a radial-petal "parterre for flowers in small beds" (rather in the manner of

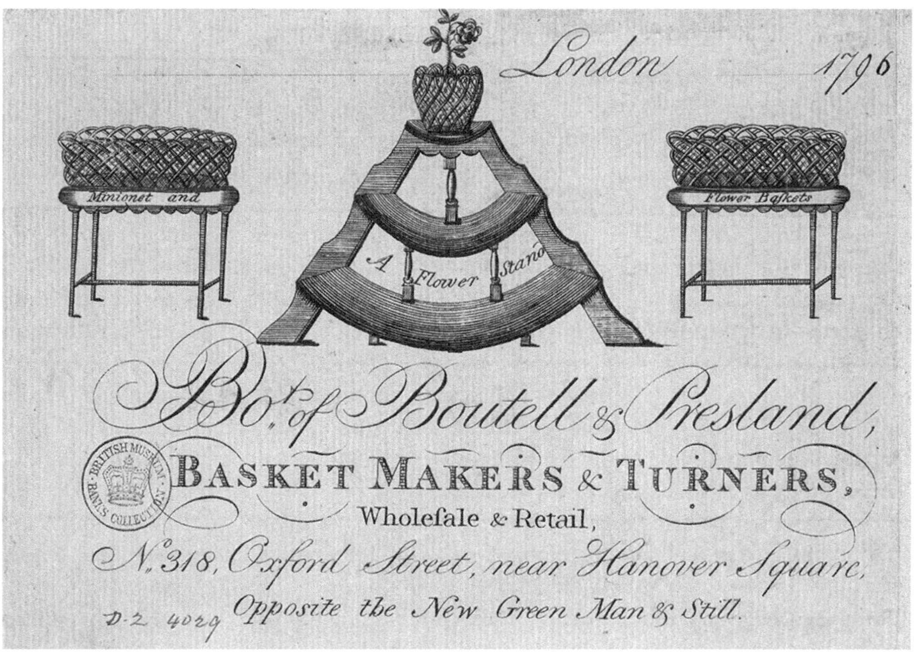

FIGURE 7.6 Tradesman's card with baskets for mignonettes and flower, and a flower stand, 1796.

Brown and Chambers), a feature opposite the greenhouse labeled "corbeille for flowers." A Red Book, consisting of watercolor drawings and detailed written instructions, often bound in red morocco leather, was a proposal of improvement to a client, which did not always result in an implemented layout. And so, without additional written or pictorial evidence, it is impossible to know if Repton's *corbeille* for Lady Wake incorporated elaborate devices in trelliswork. However, as he spent much time in the Netherlands as a youth, it seems quite probable that he would have replicated the natural simplicity of wicker baskets in the Low Countries as depicted in the works of Ambrosius Bosschaert the Elder (1573–1621) or Jan Breughel the Elder (1568–1625) (Laird 2000: 937). His daughter Mary Dorothy Repton painted a watercolor in this style, which she, or someone else, touchingly described as "designed by Papa."

Repton's predilection for this sort of Dutch-basket interpretation is seen in a proposal for the Royal Pavilion at Brighton, England, as well as in other later projects. His eldest son, John Adey Repton, introduced a variation on the theme to England when he designed the so-called Hardenberg basket for the dowager Lady Suffield's garden at Blickling Hall in Norfolk in 1823. Based on designs of 1821–2 for one of the younger Repton's patrons on the Continent, Prince Karl August von Hardenberg (1750–1822), the wooden basket was just under 2.5 meters (over 8 ft) in diameter and just under 1 meter (over 3 ft) high. Repton intended it to be filled with roses and placed within a parterre of small flowers in radial beds edged by dwarf box (Laird 2000).

By 1816 Humphry Repton had demonstrated an inclination to use heightened artifice in some of his flower gardens, for example, in the proposal he and his son made to Lord and Lady Suffield for the greenhouse at Gunton Park, Norfolk. Here they surrounded the garden with a low lattice fence, reminiscent of basketwork, and punctuated it at the corners with tall triangular wooden stands surmounted by baskets of flowers. The four stands are derived from the *torchère* or tripod candle-stand fashionable in neoclassical interiors, often placed in the four corners of the room. Interior tripods could be used for flowers during the day and candles at night, much as fireplaces served a floral and combustible purpose seasonally.

Repton took the idea even further in one of his most fanciful horticultural conceits—*Sunshine after Rain*, which depicts a "garden room" complete with tripod stands supporting goldfish bowls and flanking a marble-topped table, amidst trellises, arbors, and wicker-edged flowerbeds. It is as though he were furnishing the garden as a banqueting room after the eighteenth-century manner. The table is decked with three baskets of flowers, but instead of vases of flowers on a stretcher tying the legs together, flowers are planted underneath as though growing in a bed. This vision of a sunny garden room looks back to the Georgian interior for inspiration but reflects the modernity of Regency taste. By this time, French doors linked indoors to outdoors, and the garden had invaded the interior. The architect and design John Buonarotti Papworth (1775–1847) described a house "decorated with trellising, composed of light lath and wicker basket-work ... flower-stands and festoons" (Batey 1995: 22–3).

Humphry Repton's *Red Book for the Royal Pavilion, Brighton*, 1806, while summoning up Mughal palace and Hindu temple for his architecture of iron and glass, looked elsewhere for floral inspiration. In addition to repeating the motif of baskets of flowers, Repton called upon Francis Bacon's idea of a garden for every month: "a perpetual garden, enriched with the production of every climate" (Laird 2013b: 70–1). Through the use of purple stained glass, he even toyed with the optical illusion of transforming summer greenness into the whiteness of winter. Yet it is his shift towards flatness in flower display in the designs of 1805–15 that marks an enduring and radical departure from the "theatrical" and "mixed" style of the eighteenth century. It anticipates the "massed" bedding of the nineteenth century. The Ashridge Red Book of 1813 exemplifies this change. The emphasis on refashioning the parterre and on massing the new repeat-flowering roses in pink and red—the children's garden at Endsleigh of 1814 or the Ashridge Rosary as publicized in his *Fragments* of 1816—perhaps induced him to think in flat blocks of uniform color (Laird 2013b: 72; 2019: 39–56).

THE ENGLISH STYLE AND THE ENLIGHTENMENT IN CONTINENTAL EUROPE

In his *Andeutungen über Landschaftsgärtnerei* of 1834, Prince Pückler-Muskau (1785–1871) provided an overview of features of the English planting style in the ornamented pleasure ground, notably the ornamental "clump" or "shrubbery." One of his illustrations represented the traditional eighteenth-century style of planting continuous shrubbery with clumps, while the other depicted the new method favored after 1815 by John Nash (Laird 1999: 266–7; 2013b: 79). Pückler's published planting plan followed the Nash model for shrubbery in St. James's Park, while the Prince's novel concept of *Blumenteppiche* (flower carpets) took further what Repton was edging towards at Brighton: the flat patterns of massed flowers. Others would follow in England. For example, by 1827, at Dropmore, Buckinghamshire, the French-style parterre had returned. Each bed contained a single color of flower, massed for effect.

Although the use of plants as natural ornaments in the Regency pointed the way to Victorian artifice, English gardeners also kept up many of the practices that, for a century and a half, had been artfully balanced between the regular and irregular. They shared a common ground with the practices of architecture and the decorative arts, which involved a Pan-European interchange. Under enlightened princes on the Continent, as the English gardening style gained sway over the French in the last quarter of the eighteenth century, a synthesis arose in the decorative plantings of the pleasure ground. Two publications on the practice of planting serve to illustrate how the "natural" English style was interpreted in new mixed irregular and regular ways: *Versuch einer Anleitung zur Anlegung eines Gartens im englischen Geschmack* (Leipzig, 1794) and *Myśli różne o sposobie zakładania ogrodów* (Warsaw, 1805) (Wimmer 2001: 123–6, 134–7). The first was the work of Johann Georg Schoch (1758–1826), the son of and successor to the man who had helped plant the landscape gardens of Wörlitz for the enlightened Prince Leopold III Friedrich Franz of Anhalt-Dessau (1740–1817), Germany. The second was the creation of Izabela Czartoryska (1746–1835), who, being married to Prince Adam Czartoryski, had opportunity to landscape large estates and pleasure grounds and who went on to write a work on the Enlightenment and the Polish peasantry.

The "Garden Kingdom of Dessau-Wörlitz" is today a world heritage site, recognized as an exceptional example of landscape design and planning in the Age of Enlightenment. As a vision of estate management, it embodied scientific farming practices, technological advances, and progressive attitudes to education and religious tolerance. J.G. Schoch's 1794 publication on the layout of gardens in the English style was the first attempt by a German gardener to produce a practical work on the use of trees and shrubs in the form of "Klumps" and "Schrubs" (Laird et al. 2020). It can be taken as exemplary. This is not least because it set down detailed, illustrated specifications. His planting of evergreen shrubbery (see Figure 7.7) and deciduous shrubbery was doubtless closely modeled on the

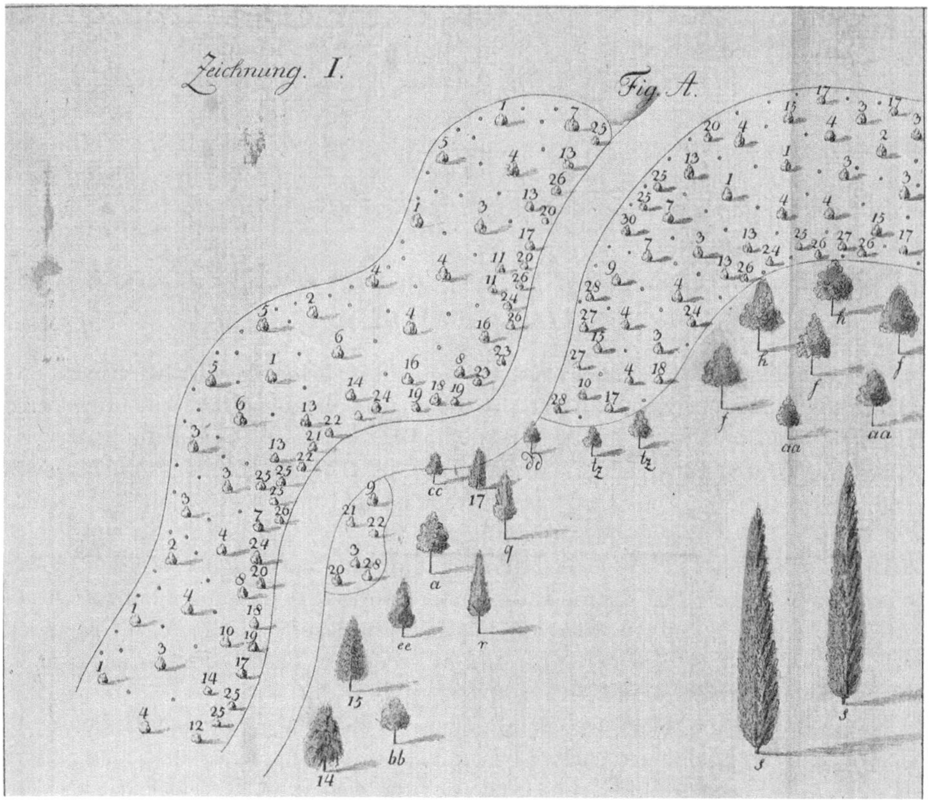

FIGURE 7.7 Detail from an illustration in J.G.G. Schoch's *Versuch einer Anleitung zu Anlegung eines Gartens im englischen Geschmack* (Leipzig, 1794). "*Zeichnung I Fig A*" shows an evergreen shrubbery, with some deciduous specimens on lawn offering lighter shades to the darker background. The numbers refer to the evergreen plant list in Chapter IV, while the letters refer to deciduous species in that chapter. Bayerische Staatsbibliothek München.

mature works of "Capability" Brown, for which we have today no equivalent planting plans, whether archived or published.

Schoch made the point that shrubs should be planted on little mounds. When such mounds, a few feet high, were extended in a serpentine band up to 15 feet wide, those serpentine rows of graduated trees, shrubs, and flowers formed a shrubbery or "Schrub." These differed from clumps or "Klumps," which were in a round shape rather than elongated like shrubberies. Within a radius of 10, 12, or 20 feet or more, there would be room to plant some taller trees at the center, surrounded by smaller trees, which would be encircled, in turn, by the shrubs and then by the perennials and annuals on the margins next to lawn. A smaller clump could consist of a single tree encircled by roses and flowers—in effect like a "bouquet."

The precise methods of planting are as important as the selection of plants. For example, because the shrubbery would appear empty at the outset, Schoch recommended placing expendable conifers in the gaps between the trees. In time these quick growers—what we might call "nurse" species—would be removed or replaced. A spacing of trees

8 foot apart in a single zigzag and the planting of narrow shrubberies would mean that plenty of light would reach the shrubs. These were mostly grouped in twos or threes. Only a special shrub—what English authors called a "specious" shrub—might present itself as a single specimen. Between the shrubs, periwinkle, ferns, and *Hypericum calycinum* could serve as a groundcover. Among the taller and lower shrubs, a few North American species introduced to England in the period of Atlantic colonies and in the Age of Linnaeus provided variety and even a little autumn color: *Physocarpus opulifolius* or *Ptelea trifoliata*; *Ceanothus americanus* or *Rubus odoratus*. Among the American perennials listed by Schoch were species of *Phlox*, *Rudbeckia*, and *Coreopsis*.

During the Enlightenment, the regions that supplied "exotics" to European nurseries and gardens included not only North America (the British and French colonies), but also the Middle East, India, and South Africa, then later, China, Australia, and South America. As just one example, a relative of today's "Butterfly Bush" (*Buddleja davidii*, introduced to Kew from China in 1896) reached Britain in 1774: the buddleia of South America (*Buddleja globosa*) (Symes 2013: 74–84, 86–7). It can be argued that a particular impetus for the graduated shrubbery of Schoch's account was the influx of North American species, which required a feature of display derived from "theaters of flowers" but reflective of Linnaean ordering of plants into a classificatory system known as the "sexual system." Linnaeus' milestone in Enlightenment thinking helped order the shrubbery as a scientific collection as well as an aesthetic assemblage.

It was an Irish gardener, Denis McClair, who went to Poland in 1790 at the request of Princess Izabela Czartoryska. She was already employing an English gardener and designer, James Savage, at her estate of Puławy near Lublin. In time, McClair would be responsible for sending the scented *Rhododendron luteum* to England to augment the American azaleas and ahead of the influx of Asian species of *Rhododendron* (Symes 2013: 92). Czartoryska's book was entitled "some thoughts on the laying out of gardens." Her purpose was to introduce the English style to Poland, thus favoring an imitation of nature over art.

Clumps (*klómby*) were given special attention in her work, and, through several expressive illustrations, the naturalness of tree and shrub forms made a striking counterpoint to the regularity of the plan form, which was often elliptical (see Figure 7.8). Her clumps were composed of trees, shrubs, and flowers, all arranged in graduation. As just one example, the center of this particular ellipse would be filled with a mixture of evergreen and deciduous trees: from *Robinia pseudoacacia* to fir, birch, silver poplar, Lombardy poplar, larch, spruce, field maple, ash, and elm. The lower-growing trees would include rowan, bird cherry, lilac, and elder. Among the taller shrubs were *Viburnum lantana*, *Cornus sanguinea*, *Robinia hispida*, Persian lilac (*Syringa* x *persica*), broom, and snowball tree (*Viburnum opulus* 'Roseum'); among the lower were *Spiraea* spp., *Rubus odoratus*, roses, and *Daphne* spp. Just inside the lowest edging of flowers, there was a row of taller flowers intermixed with roses and *Prunus tenella*, a double-flowered dwarf cherry.

It is her vision of a single flowerbed that stands out among the "clumps"—a jewel of a *klómb* (Laird 1999: fig. 223). The representation is somewhat deceptive in showing the crown imperial (*Fritillaria imperialis*) of spring in flower during the climax of summer blooming (the hollyhock that forms the pinnacle alongside the rose bush would only achieve its full flowering in mid-summer, when the crown imperial was long faded). The use, however, of many annuals—French marigold, candytuft, marigold, nasturtium, double balsam, and Venus' looking glass—allowed quick-growing flowers to fill such

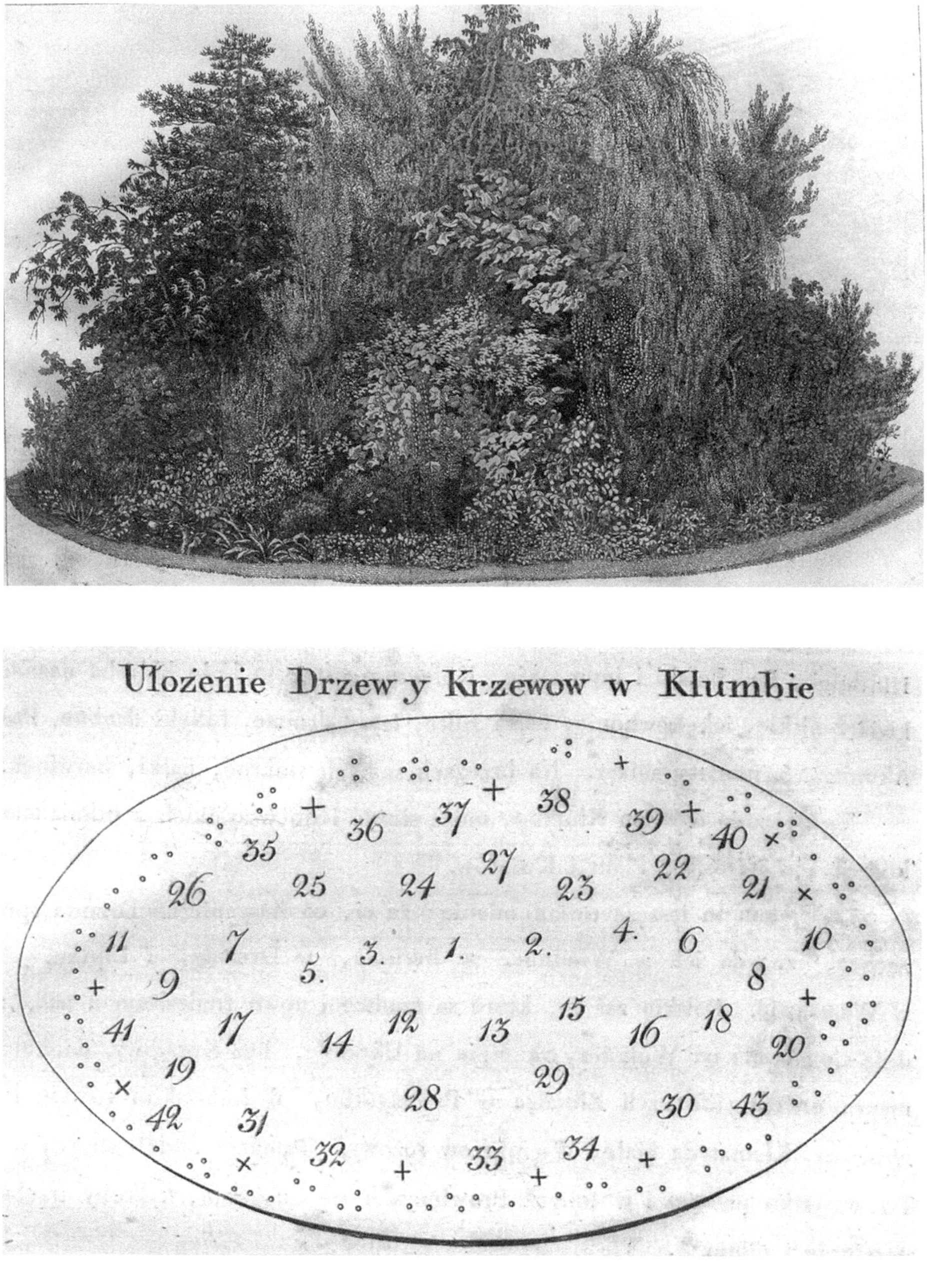

FIGURE 7.8a and b Planting plan and illustration of a clump in Izabela Czartoryska, *Myśli różne o sposobie zakładania ogrodów* (Warsaw: Biblioteka Główna Politechniki Warszawskiej, 1805; 3rd edn. 1808).

fading vacancies. American perennials—*Phlox paniculata*, *Coreopsis verticillata*, and *Monarda didyma*—made significant splashes of mauve, yellow, and red in the middling range, set off by the evergreen conifer *Juniperus sabina*. Biennials were represented by the medium-high foxgloves and Canterbury bells, as well as by tall hollyhocks. Around the margins, the lower-growing annuals were complemented by the evergreen periwinkle and European wild flowers such as sneezewort (*Achillea ptarmica*) and woody nightshade (*Solanum dulcamara*). This fanciful assortment of flowers is slightly puzzling in terms of practice, but the "theatrical" structure has the unmistakable authority of English models brought to perfection by Lady Elizabeth Lee in her 1799 planting plans for Hartwell House, Buckinghamshire (Laird 1999: 363–76).

WOMEN AND NATURAL HISTORY AND GARDENING IN THE AGE OF ENLIGHTENMENT

It would be misleading to argue that technological advances (other than developments in hothouses, first, and glasshouses, second) shaped major changes in gardening in the period up to 1815 (for glasshouse technology, see Simo 1988: 111–16).

By contrast, Linnaean botany had a profound impact on local collecting, global plant hunting, the nursery trade, and women's engagement with natural history, botanical art, and the accomplishments of gardening. The fact that Izabela Czartoryska's model for a flower *klómb* had Latin binomials specified (e.g. *Phlox paniculata* rather than Lady Elizabeth Lee's outdated Latin "Lichnidea," which was embedded amid her vernacular names) points to book learning. Czartoryska's symbols for perennials, biennials, and annuals are also taken from a scientific horticultural literature (Aiton 1789: xxx).

Well before Linnaeus, the 1st Duchess of Beaufort had been engaged in a variety of horticultural and botanical pursuits (Laird 2015: ch. 2). These led to her florilegium of 1703–5, painted by the German artist Everhard Kick (known as Kickius), and her herbarium of pressed plants, assembled with the aid of Sir Hans Sloane (1660–1753). Concluded on her death in 1715, it would be absorbed into the Sloane Herbarium, now at the Natural History Museum, London. Collecting and displaying florists' flowers—from auriculas to carnations—had been her passion at Beaufort House, Chelsea, until the early 1690s, when an interest in the hothouse technologies of Hampton Court made her into an avid collector of tender exotics. While she never carried through a wholesale conversion of her conservatory into a tropical stove, she was preoccupied with naming and picturing the products of tropical Asia and Central America, along with succulents of South Africa and many temperate species.

The composition of folio 15 of Kickius' florilegium suggests a group of plants chosen explicitly to display contrasts of foliage (since their flowers are insignificant). It is reasonable to suggest, then, that, aside from the science of polynomials, referenced through botanical works (e.g. those of Leonard Plukenet (1642–1706), Queen's Botanist at Hampton Court), the art of plants as natural ornaments was at issue. Broad-leaved butcher's broom (*Ruscus hypophyllum*), known to gardeners since 1625, was depicted in an unlikely, but exquisite, trinity with the feathery tropical "Asparagus Zeylanica &c." (as yet unidentified) and "Phyllitis seu ling: cervina minor crispa folio Pluk: 248: 2" (native hart's-tongue fern, *Asplenium scolopendrium* var. *cristatum*). The duchess' interests in the aesthetics achievable in horticulture were thus, perhaps, being tested out in Kickius' compositions.

By 1799 Kickius' globe amaranth (*Gomphrena globosa*), and convolvulus (*Ipomoea hederacea* and *Ipomoea coccinea*) were all in Lady Elizabeth Lee's flowerbeds at Hartwell House. Indeed, the staking of the "Convolvulus" on folio 30 of the Kickius florilegium remained the practice one hundred years later. Plants in the Hartwell plans also appear in the Duchess of Beaufort's *hortus siccus*. William Aiton (1731–93) documented in his

Hortus Kewensis of 1789 that oriental persicaria (*Persicaria orientalis*, as *Polygonum orientale*), a sunflower (*Helianthus giganteus*), and Culver's root (*Veronicastrum virginicum*) were first cultivated by the duchess. The two giants from West and East—the perennial sunflower and nodding persicaria—turned Lady Elizabeth Lee's beds into late-blooming wonders full of flowers to brighten the autumn.

William Aiton's *Hortus Kewensis* of 1789 (with 2nd edition of 1813 by his son) had begun life as a work of the same title by physician and naturalist John Hill (1714–75), and in association with John Stuart, 3rd Earl of Bute (1713–92). Bute was the primary force behind the creation of the scientific grounds at Kew. Princess Augusta's engagement with Lord Bute in turning a "Desart" into "Eden" at Kew was far from scientific, yet no less significant in terms of Enlightenment sensibilities (Laird 2015: ch. 5; 2017a: 140–4). Her floricultural arena—the Aviary Flower Garden—was what William Chambers had laid out in petal-like radial symmetry. But her impress came through the sonorous, refulgent, and perfumed conjunction of English songbirds with the products of English flower craft: the double hyacinths and carnations of spring and summer display.

A synthesis of sound and scent was something enjoyed by men and women, especially those with a literary bent, as the letters of Horace Walpole and Jemima, Marchioness Grey (1722–97), confirm. Yet, in often spending more time than men in the garden (and in sometimes outliving them), women had a particular claim to this affective epistolary language. On June 7, 1747, arriving in the pouring rain, Jemima was deprived of her greatest pleasure amid other delights: "The Evening was not fine enough for the Nightingale to favour us with his Song, but the Blackbird has scarcely ceased, & a Cuckoo welcom'd me .... The verdure delightful, & some of the Walks quite prefum'd with Honeysuckles & Ceringo's." The following year, on June 2, 1748, the sight of honeysuckles "hanging in finer Festoons that Art could imitate among the Bushes" led her to contemplate a circle of her friends making art out of nature with "their best-tipp'd Pencils" (Laird 2017b).[1]

Mary Delany used the same epistolary language when writing about her garden at Delville on the edge of Dublin on March 29, 1746: "Our garden is now a wilderness of sweets. The violets, sweet briar, and primroses perfume the air, and the thrushes are full of melody and make our concert complete" (Laird and Weisberg-Roberts 2009: 151). She employed her own sharpened pencil to create numerous sketches of her garden and Irish environs.

What was, early on, Mary's versatility in flower-based accomplishments, most notably in the design of her flower-strewn silk court dress of 1740–1, became in later widowhood the prodigious production of nearly a thousand botanical collages. Delany would refashion herself to undertake a pious interpretation of an Enlightenment enterprise, a ten-year devotion to her *hortus siccus* or "Flora Delanica": first, by benefiting, after 1768, from perennial exposure to Bulstrode as the home of the widowed amateur naturalist, the 2nd Duchess of Portland (1715–85); second, by learning the Linnaean sexual system through the dissections and compositions of the greatest botanical painter of the age, Georg Dionysius Ehret (1708–70); and third, by absorbing Linnaean classifications of English plants in her annotated English transcription of William Hudson's *Flora Anglica* (1762), and with the aid of the Duchess' "botanical master," the Revd John Lightfoot (1735–88).

In organizing her collages as an "album amicorum" in dictionary format, she did not follow a scientific classificatory order (Campbell Orr 2019: 294–9). A woman of letters, it would have come naturally to her to make the alphabetical ordering a means for her and her circle of friends to enjoy the collages in leisured sociability rather than as students of Linnaeus per se. Yet her decision to make Latin binomials rather than common names the basis of the alphabetical order likely reflected the way that she and the duchess would have thought of plants, whether as natural ornaments in cultivated compartments in the garden, whether as "natural embroidery" out in the countryside, or whether as cut flowers in a vase. Linnaeus' "artificial"[2] system had become second nature to them in the artifice of using plants to ornament life.

CONCLUDING REMARKS

Since Europe was part of a global exchange network, it is important to conclude this chapter with a brief mention of non-European gardening beyond the Enlightenment. For example, the tulip, having been first introduced to Europe from Ottoman Turkey in the sixteenth century, gave rise to "Tulipomania" by the mid-seventeenth century. Hence any discussion of florists' flowers or flowers of distinction, linked to early "theaters of flowers" (as discussed above), appears incomplete without recognition of the way the Ottomans gardened in traditions influenced by Roman, Byzantine, and Islamic gardens (Atasoy 2007). The Ottomans developed their own exquisite sensibility in floricultural artifice and sensitivity to natural settings in the embrace of geometry. Exchange of technical know-how provides a second example of global networks and the Enlightenment. Grafting, developed by the Greeks and Romans, paralleled grafting techniques in ancient China (Métailié 2007). Thus, when, amid an influx of Chinese plants reaching Europe around 1800, tree peonies *mudan* (*Paeonia suffruticosa*) came to be cultivated as European cultivars (e.g. 'Joseph Banks'), it was on the basis of centuries of Chinese experimentation with stock and scion. *Rosa banksiae* 'Lutea' followed thereafter, and in 1819 John Reeves, a tea inspector for the East India Company, imported tea roses from China, which were valued for indoor display in pots, or sold in Parisian markets "enveloped in coloured paper" (Blacker 2000: 133).

In the Enlightenment, therefore, European ways of using plants as natural ornaments benefitted from a cultural hybridization, notably through trade. The period saw the rise of national trading companies, for example the East India, Hudson's Bay, Turkey and Russia Companies, and rivals such as the Dutch East India Company or French East India Company (Sloan 2003: 231). It would be impossible to overstate the importance of the Chinese export trade in ceramics during the Enlightenment (Battie 1990: 49). The tin-glaze earthenware (Delft), which was made in England before and after the English began to make their own porcelain in the mid-1740s, sought to reproduce the Chinese blue-and-white patterns that had monopolized the market within the general fashion for Chinoiserie. Thus, around 1700, Queen Mary might have had a Delft urn filled with sunflowers, red opium poppies, and African marigolds amid foliage of ivy and tulip tree. By 1800, flowers, arranged in a Chinese export tobacco-leaf pattern *seau*, could include agapanthus, acanthus, lilies, carnations, tulips, and chrysanthemum representing, by origin, the continents of Africa and Europe and cultures from the Middle East to the Far East (Blacker 2000: 53, 78).

While the architect John Vanbrugh (1664–1726) lived and worked in India between 1683 and 1685 and drew inspiration therefrom (Williams 2000), William Chambers used first-hand experience of the gardens of Canton to inform his vision for Kew. Just as "India" inspired Repton's proposals for Brighton, so too "China" is behind much that Repton sketched and wrote about in his Red Book for Woburn Abbey. In a similar second- or third-hand way, Chambers' contemporary, the topographical/botanical artist Thomas

Robins the Elder (*c.* 1716–70), who never set foot outside England, brought a Rococo-Chinoiserie artistic sensibility, derived from his work as a fan-painter, to his portraits of English plants and English plantings (Laird 2015). With both spontineity and finesse, his paintings naturalized exotics, while his natives were correspondingly exoticized. Framing his art around an ornamentated milieu of places, plants, and other living things, Robins imaginatively melded four continents—Africa, Asia, America, and Europe—into a singular vision, but notably a vision that is hard to fit into received Englightenment histories of natural history and gardening in the manner of Mary Delany and the Duchess of Portland (note his absence in Sloan 2003 and in Bending 2013a). For plants as natural ornaments in global terms, another history is needed.

CHAPTER EIGHT

The Representation of Plants

EKATERINA HEATH AND JENNIFER MILAM

Between 1650 and 1800, the representation of plants in art shifted from the inclusion of religious references to the stimulation of secular reflections. In medieval and Renaissance art, plants were largely included as didactic symbols tied to specific meanings outlined in the Bible and emblem books, which combine allegorical illustrations with explanatory text (Woldbye 1991: 46). From the mid-seventeenth century onward, plants were less often infused with narrative purpose. This change was driven in part by a new skepticism towards religion triggered by philosophical debates. In this regard, several aspects of Enlightenment thought had implications for the visual representation of plants during this period: the philosophical contemplation of a vegetal "soul"; the poetic endowment of noble status upon trees; the development of a science of botany, informed by voyages of discovery and exploration; and theories of aesthetics that revolved around a notion of beauty observed in nature, with flowers valued purely for their colors, shapes, and form. These shifts in thinking about plants were paralleled in visual representations by an artistic treatment that urged viewers to contemplate plants for reasons other than interpretive readings based on inherited symbolism, even while such meanings continued to circulate and be adapted to social and historical changes. When considering the visual representation of plants, it is therefore useful to explore not only what plants mean in the context of a particular subject matter, but also how the plants were handled visually to stimulate responses tied to and informed by Enlightenment thinking about nature.

VEGETAL SOUL AND NOBLE TREES

Toward the end of the seventeenth century, philosophers pondered the possibility of a vegetal soul. In 1704 the prominent German rationalist Gottfried Leibnitz (1646–1716) argued: "The great analogy which exists between plants and animals inclines me to believe that there is some perception and appetite even in plants; and if there is a vegetal soul, as is generally thought, then it must have perception" (Leibnitz 1996: 139; Marder 2014: ch. 7). This proposition influenced a more considered attitude towards plants, especially those that were perceived to have ancient associations, such as oak trees. As subjects of representation, plants were increasingly pursued as objects of interest and beauty. Eighteenth-century writers endowed trees with sublime dignity and artists portrayed notable examples (Thomas 1983).

The English poet Alexander Pope (1688–1744) claimed that a tree was "a nobler object than a prince in his coronation robes" (Pope 2006: 574), while William Gilipin (1724–1804), the artist and writer credited with originating the idea of the picturesque, published books like *Remarks of Forest Scenery* (1791) to encourage readers to look at trees and woods in their local landscapes as artful subjects in their own right. Prints like *East View of Yardly Oak* (see Figure 8.1) by William Cowper (1731–1800) include details of the age, size, and ownership of particularly prominent examples, and included poetic verses to stimulate aesthetic contemplation and noble meanings: "Time made thee what thou were—King of the woods" (Cowper 1808: 167). Following the conventions of seventeenth- and eighteenth-century landscape painting, an image that was primarily concerned with capturing a picturesque view would include the largest tree on the right- or left-hand side of the vista as a framing device. "Yardly Oak" is instead placed at the center of the image, assuming prominence as the primary object of the beholder's attention. More than representations of landscapes, these types of pictures are essentially portraits of trees that endow plants with a living presence and transcendental soul.

This sensibility enhanced an analogy that developed during the period between great trees and great families, especially in Britain. The Irish philosopher Edmund Burke (1729–97), for example, described the aristocracy as "the great oaks that shade a country" (Burke 1984: 184). Prominent oaks on the landed estates of the English gentry assumed familial meaning, which artists like Thomas Gainsborough (1727–88) incorporated into paintings that further blurred the lines between still life, landscape, and portraiture. In Gainsborough's *Mr and Mrs Andrews* (*c.* 1750; National Gallery, London) the oak is as central to the group portrait as the human sitters, symbolizing the

FIGURE 8.1 Walden Henry Hanmer and John Laporte, *The Cowthorpe Oak, Yorkshire* (1806). Hand-colored etching. British Library (Maps K. Top.45.41.2).

continuity of generations on the family estate. With the couple posed in front of this great oak, Gainsborough takes a wide view of The Auberies, their nearly 3,000-acre estate in the Essex countryside, balanced on the opposite side of the composition by a fertile field of wheat, and completed in the background by a well-stocked pasture on the right and unpruned trees on the left, denoting the woodlands of the estate. The free growth of these trees in the landscape conveyed the Englishman's political and social freedom, in comparison with their courtly European peers. The Scottish editor of John Evelyn's seventeenth-century text *Sylva, or A Discourse of Forest-Trees*, Dr. Alexander Hunter (1729–1809), summed up this eighteenth-century taste for uncultivated trees in 1776, writing: "Everyone who has the least pretension to taste must always prefer a tree in its natural growth" (Evelyn 1776: 379n). Trees became so personal to families by the end of the eighteenth century that the English philosopher and social reformer Jeremy Bentham (1748–1832) proposed country gentlemen should embalm their ancestors and place them in avenues, alternating them with trees (Bentham 2002: 3a). Oak, in particular, symbolized masculinity, vigor, strength, and stability in representations (Stafford 2016: 98). The English poet and early landscape-gardener William Shenstone (1714–63) wrote that the oak was "the perfect image of the manly character ... the British one" (Shenstone 1764: 134).

European oaks were not the only subjects to occupy the artist's attention. Particularly remarkable elm trees in America and banyan trees (*Ficus benghalensis*) in India were portrayed as guardian spirits in the foreground of landscapes that both dwarf and complement churches and temples that appear in the distant background of images created by early travelers and colonists (Payne 2017). Functioning less as symbols than as metaphors guiding contemplation, such images of plants lead viewers to reflect on the ancient history of primordial nature in tandem with a developing narrative of man's progress.

SCIENCE OF BOTANY AND VOYAGES OF DISCOVERY

In the long eighteenth century, images of exotic trees and plants acquired an added significance as symbols of the imperial aspirations of European countries. During voyages of discovery, visual representations were the preeminent means of knowledge production and dissemination, replacing textual descriptions and natural-history specimens (Bleichmar 2012: 3). Images of plants were aimed at making imperial ambitions visible, in order to gain knowledge about faraway places that would be used to benefit and grow a nation's empire. This attitude led to the creation of a Pan-European visual culture of botanical illustration, which can be characterized by its extractive nature. Plants in botanical representations were typically depicted against a white background which erased any references to the plant's natural habitat. These images mirrored the objectives of colonial economic botany—the practice of removing species from their territories and transplanting them to new locations, where they would benefit the empire.

Before the seventeenth century, botanical literature primarily consisted of herbals in which the selection of plants and their descriptions were determined by their medicinal use. During the 1600s this literature transformed into botanical work in the modern sense, whereby descriptions of plants and their placement in relation to each another depended more on the taxonomic relationship between them than on their uses. A clear division into categories of medical, botanical, and horticultural literature was, in general, established with an expansion of production owing to this variety of purpose and increase in popularity. Botanists began to travel to remote destinations, and technical advances in printmaking supported an improved accuracy of representation. Illustrations of plants, which previously had been supplementary and non-essential to the text, became the key elements of botanical publications.

Copper engravings largely replaced woodcuts in the course of the seventeenth century because they were more suitable for reproducing details. Woodcuts continued to be used for botanical images due to their low price, but more precise engraving techniques and etching on copper revolutionized the art of botanical illustration. A skilled engraver could employ the etching needle to capture light and shadow, with the burin (engraving tool) creating remarkably sensitive lines and chiaroscuro effects. These techniques were highly suited to rendering plastic values and three-dimensional form. Many scientists took advantage of this technique to reproduce fine morphological details and to capture delicate hues. At the same time, the realism of these images was enhanced by hand-tinting with gouache (opaque body colors).

Artists continued to experiment with print-making techniques throughout the eighteenth century to obtain even more subtle effects. Color-printed mezzotints became available in this period, but they were less suited to scientific illustration because the technique yielded slightly smudged images. Aquatint produced more precise images, and therefore improved accuracy in the representations of plant forms.[1] In spite of these technical developments that enhanced the accuracy of representation and served scientific study, the aesthetic appreciation of plants in art continued. Towards the end of the eighteenth century, French botanical illustration merged these parallel impulses of scientific illustration and artistic value. A professor of flower painting at the Jardin des Plants, Gerard van Spaendonck (1746–1822), and his student, the artist and botanist Pierre-Joseph Redouté (1759–1840), both experimented with stipple-engraving techniques. This method was based on distributing a pattern of dots on the copper plate, making it possible to reproduce the most minute details of color and texture of plants (Tomasi 1997: lv).

Hortus (Indicus) Malabaricus (1678–93) was the first significant book published on the medicinal properties of Indian plants. Consisting of twelve volumes with 794 copper engravings, the project was conceived by Hendrik van Rheede tot Drakenstein (1636–91), the Governor of Dutch Malabar. As a prime example of botanical illustration in the seventeenth century, such images were limited to black-and-white reproduction. Copper engravings allowed publishers to produce books relatively cheaply, but valuable information was omitted because color was absent. Images here captured the plants' natural environment and their relative scale by adding depictions of the local people. Each image combines several views of the same plant without a coherent approach to representation. In some instances, the artist represented flowering plants within a landscape; in others the artist's focus was on a plant's leaves. In the absence of any unifying theory of taxonomy, the artists depicted a plant's most distinctive features, be it size, the distinguishing shapes of leaves, or remarkable flowers. Roots and seeds appear sporadically. Overall, the representation of plants in *Hortus Malabaricus* sought to enchant viewers with the exotic nature of Malabar. Idealized and spectacular, the imagery promoted the wealth of the Dutch East India company and its access to useful plants (Heniger 1986: 95, 124–38; Spudich 2008: 45–61).

Representations of plants within *Hortus Malabaricus* contrast sharply with eighteenth-century botanical images produced after the British voyages of discovery in which there are rarely any references to local landscapes and population. These later images became standard in their focus on the floral parts of each plant, thereby often omitting its most characteristic features. *Phormium tenax* (or New Zealand Flax) (see Figure 8.2) by

FIGURE 8.2 Sydney Parkinson, *Phormium tenax*, 1775. The Natural History Museum, London. © Alamy.

Sydney Parkinson (?1737–71) epitomizes the expression of economic botany in visual form. Commissioned by Joseph Banks (1743–1820) and sketched during the *Endeavour* voyage of 1768–71, the image transformed a local specimen into a decontextualized plant that could function in the global economy. Capturing plants was part of a strategy aimed at seeing, knowing, and ruling the empire. It was based on an extractive view of nature and the pursuit of imperialism by European nations (Bleichmar 2012: 79 and 151).

Late-eighteenth-century botanical illustrations were heavily influenced by Linnaean botany, which, in part, erased the environmental and cultural contexts of plants. The white background and snipped twig in images like Parkinson's *Phormium tenax* visualized the idea that a plant could be extracted from one natural environment and inserted into another with ease. As a result of illustrative processes, a plant acquired an identity separate from what it had in its original environment. The standard types of images and omission of background allowed a student of natural history to compare specimens from all over the world in their study or library. Scholars examined plants from what might be termed "a global perspective," perceived as objective and permanent. Such images were effective for classification, but conveyed little in the way of additional information to deduce the plant's cultural meaning or local uses (Tobin 1996: 259).

Parkinson's watercolor presents the plant at its peak, vibrant and flowering. The taxonomic system developed by Carl Linnaeus (1707–78) brought a revolutionary change to botanical illustration, focusing the viewer's attention on flowers, whereas earlier approaches had often shown the whole plant, including roots, stems, and seeds. In spite of the fact that the value of New Zealand flax lay in its strong fibrous leaves, the bulk of the image is occupied by flowers with prominently displayed pistils. Such an emphasis was needed for ease of identification; yet the focus on flowers encourages enjoyment of the plant as an object of beauty rather than something meant to be used in the production of fibers.

The visual representation neglects to relay the plant's origin, local uses, cultural meaning, or lifecycle; instead, it is an idealized and generalized view of this species' appearance, achieved through the observation of multiple plants of the same type and combining these observations into one image. Parkinson created an outline drawing of the plant by studying two cuttings, which are preserved as an herbarium sheet in the Te Papa museum.[2] Analysis of this herbarium sheet confirms that Parkinson's watercolor is not an exact representation of any one of these cuttings, but rather a combination of the two. The drawing lacks specific features of the plant as found in nature—blemishes, uneven shape or color, traces of insects, or dried leaves. Parkinson introduced deliberate distortions to assist with identification: the pistils are arranged in even rows, ready for the botanist to count, and a leaf is added to the flowering shoot, when in fact leaves never grow so close to the flowers on this particular plant. To enhance the visual pleasure of the image, the composition and use of proportions draw the viewer's eye away from the leaf to focus on the delicately colored flowers. The experience is one that combines botanical identification with aesthetic appreciation.[3]

Paradoxically, Parkinson's depiction of New Zealand flax, which for Banks remained the most economically valuable plant that his team collected on his voyage around the world, communicated none of the benefits derived from the transformation of its leathery leaves into useful and hard-wearing materials. Rather, the representation of the plant focused on its flowers, aesthetically pleasing, and necessary for correct identification and classification. The image's usefulness in its promotion of economic botany lay in the fact that its white background erased the plants' local uses and the viewer was shown that it was ready for extraction and replanting in a different part of the world deemed suitable in London. A full understanding of the plant's benefit could only be gained by reading the accompanying text and cross-referencing it with other plants depicted in a series.

With increased contact between China and the West in this era of expanded global trade routes, exploration, and cultural exchange, Chinese representations of flowers changed to include Western elements during the eighteenth century. Giuseppe Castiglione (1688–1766), a Jesuit artist working for the Qing emperors between 1715 and 1766, initiated an artistic form that sought to blend aspects of traditional Chinese flower painting with Western forms of still life and botanical illustration. Castiglione's *Vase and Flowers* exemplifies this novel approach, which fused Western techniques, such as chiaroscuro, with Chinese methods and materials (see Figure 8.3). Three generations of Qing emperors responded to Castiglione's style of realism and increased visual impact with encouragement for this cross-cultural practice of plant representation that enhanced their imperialist agenda.

In Castiglione's painting, a double-headed peony sits in a Ming-style vase as a symbol of the submission of supporters of the Ming Dynasty to the Qing emperor. Peonies, or "The Kings of Flowers," came from the rebel Jiangnan province, which continued to struggle against Qing rule for many years after its defeat. Commonly grown in Jiangnan gardens, the flower became a symbol of this territory. Castiglione extracted the flower out of the garden context and placed it in the format of a tribute painting, aimed at recording the taxpayers' gifts to the court. Captured against the plain white background, like its European counterparts, this plant reinforced the notion of imperialism and domination. Verdant green leaves of the peony signify the promise of growth and abundance, communicating the benefits of submission to the Qing emperor's will. The artist's employment of European methods of modeling communicated the global outlook of the Qing emperors, who were interested in promoting their successes to the West. Furthermore, by incorporating Western techniques into Eastern forms of flower painting, Castiglione's scroll reinforced the Qing Dynasty's ultimate vision of Beijing as the center of the world (Chiem 2018).

GENDERING REPRESENTATION AND THE SEXUAL NATURE OF PLANTS

While botanical representation was enhanced by technical innovations and aesthetic purpose in the eighteenth century, the development of scientific botany emphasized the sexual nature of plants. An increased emphasis on the Linnaean taxonomy, based on the number of sexual elements of the flowers, transformed the look of botanical studies by emphasizing their pistils and stamens. This recognition by artists of flowers as sexual organs contributed an additional layer of meaning to the images they produced and drew distinctions between appropriate and inappropriate manners of depiction. Publications aimed at amateur female artists, like Augustin Henckle's *The Lady's Drawing Book* (1753), specifically avoided drawing attention to the reproductive parts of flowers (Shteir 1996: 41) by training women to focus instead on depicting the outlines of leaves, stems, thorns, and petals.

During the period of Enlightenment women were encouraged to pursue amateur forms of art-making and scientific knowledge as polite diversions. After the 1760s, botany became a particularly fashionable pursuit for women because it was perceived as both suitable and "improving" (Shteir 1996: 2–3). According to the *philosophes*

FIGURE 8.3 Giuseppe Castiglione (Lang Shining, 1688–1766), *Vase and Flowers*. Hanging scroll, ink and color on silk, 113.4 × 59.5 cm. National Palace Museum, Taipei. © Alamy.

of the French Enlightenment, women were closer to animals, and consequently to plants, making the study of botany appropriate for them (Spencer 2012). Others simply maintained that the study of plants was aesthetically aligned with feminine beauty. Women collected and dried plants, wrote letters to botanists, and made drawings and paintings of flowers. They were rarely involved in plant taxonomy or advancing botanical science, as their activities were relegated to the domestic sphere, their kitchens, or dining-rooms (Schiebinger 2004a: 36–7).

While botanical art was a socially acceptable activity for women, they were generally discouraged from creating scientific botanical images. The books targeting female students avoided teaching them to draw the roots and reproductive parts of the flower. Botanical art became popular among women because of its domestic associations and the ready accessibility of the subject matter. Plants were suitable for close leisurely looking without the need to worry about the cost of, and gender restrictions imposed on, using a human model. Painting plants could take place in the domestic sphere and the resulting image could be used for decorating the house with depictions of plants or later used for embroidery or other craft. Anna Maria Garthwaite (1688–1763) used realistic representations of plants for designing fashionable fabrics (Anishanslin 2016), while Mary Delany (1700–88) created "paper mosaiks" from colored tissue paper (see Figure 8.4). Her representations of plants were unprecedented in their botanical accuracy and praised by the likes of Joseph Banks who proclaimed them to be "the only imitations of nature. ... he could venture to describe botanically" (quoted in Shteir 1996: 44).

The botanical accuracy of many such works as well as the desire of some women to go beyond the prescribed behavior encouraged them to depict the roots of flowers or pistils, achieving innovative visual effects. Delany became one such artist who defied the limitations imposed upon her gender by the literature on botanical art for women. Her images focus on the reproductive parts of the flowers—pistils and a pestle. *Magnolia Grandiflora* (1776) allows the viewer to literally count every sexual part of the flower and classify the plant according to Linnaean principles. This scientific precision placed this image on the same level as botanical watercolors produced by male artists at the time. Delany's use of a black background also differentiated her collage from traditional botanical images, making the image more decorative and at the same time following the principle of extractive imagery where the specimen was depicted outside of its usual context. The choice of material for her art was significant. On one level, it was crafted, something that was not perceived to be serious and scholarly; on another level, the detailed precision of these cut-outs is so high that it creates an impression of a watercolor. This *trompe l'oeil* technique attracted the viewers' attention to admire not only the work's scientific precision but also the high degree of craftsmanship and artistry. The aestheticization of the surface of the work occurred through the highly skilled use of unusual materials. Depictions of flowers by Delany and other female artists, such as the French painter Anne Vallayer-Coster (1744–1818), have a strong sensual appeal. Delany achieved this through depicting overlapping petals that looked like female sexual organs, something Vallayer-Coster accomplished through her mastery of layering paint. The visual enticement of such images is a continuation of the tradition of seeing a life of plant as parallel to that of a human being (Moore 2005).

AESTHETICS OF STILL LIFE

Concurrent with these botanical developments which involved close looking at plants, Enlightenment philosophers across Europe began to contemplate nature for aesthetic rather than symbolic or scientific purposes. In their writings, the Englishman Joseph Addison (1672–1719) and the German Immanuel Kant (1724–1804) valued plants for their colors and shapes over meanings that derived from traditional symbolism. Addison urged readers of *The Spectator* to delight in the wild appearance of plants found in nature as objects "more qualified to entertain the imagination" than a work of art (June 25, 1712, 414), while Kant described a tulip in the *Critique of Judgement* as something "beautiful,

FIGURE 8.4 Mary Delany, *Magnolia Grandiflora* (Polyandria Polygynia), formerly in an album (Vol. VI, 57); the grand Magnolia. Collage of colored papers, with bodycolor and watercolor, on black ink background, 1776.

because we meet with a certain finality in its perception, which, in our estimate of it, is not referred to any end whatever" (Kant 1953: 80). This shift in emphasis to a focus on capturing not only the visual appeal of plants, but also their value as objects of aesthetic contemplation, can be seen in the rise of still lifes during this period, especially in the works of the eighteenth-century French painter Jean-Baptiste-Siméon Chardin (1699–1779).

Arrangements of plants and flowers were popular in Europe throughout the seventeenth and eighteenth centuries, particularly when they were handled with the technical mastery of Dutch still-life painters. The sheer skill and effort of their production conveyed the economic value of and investing in the work of art, which appealed to developing merchant-class and bourgeois audiences in the period (Bryson 1997: 107–61). While appreciated for their realism, the majority of these paintings of plants incorporated religious and allegorical symbolism that would cause viewers to reflect on their own mortality. Moreover, they were not true to nature. Unlike the portraits of trees that appeared later in the eighteenth century, seventeenth-century still lifes were not portrayals of actual bouquets or natural formations of flowers. These representations of plants often include ones which flowered in different seasons, overflowing from vases that are too small to hold the exaggerated arrangements. The labor of horticulture, including the forcing of varietals, and the painstaking effort of the painter's craftsmanship, cooperate to convey the idea that the value created by the representation of plants derives from human effort alone.

Still-life specialists like Jan Davidszoon de Heem (1606–84) subverted the principle of symmetry in their compositions as a visual cue to direct viewers towards hidden meanings (Schneider 1996). In de Heem's *Vase of Flowers* of 1660 (see Figure 8.5) the leaves surrounding the flowers have begun to wrinkle and turn inward, a sign that the wilting phase has begun. The brevity of the flower's life could be compared to a human one, both ended before fully realized. This is particularly true of flowers like peonies and anemones, which lose their petals quickly. To connect these reminders of death with religious purpose, de Heem has interspersed the flowers with ears of grain, an allusion to the Eucharist. Butterflies and snails also appear as spiritual references associated with Christ's resurrection. Seventeenth-century European still lifes, above all else, exhort the beholder to praise God and contemplate the brevity of life, death, and the hereafter (Heilmeyer 2006; Segal 1990).

As *memento mori*, still lifes of cut flowers and plants refer to the lifecycle and the inevitability of death. Despite the gravity of these reflections, still-life subjects were perceived to have the lowest status in the hierarchy of genres established by the French Academy during the second half of the seventeenth century. In 1667, André Félibien (1619–95) proclaimed in the *Conférences de l'Académie* that "He who produces perfect landscapes is above another who only produces fruit, flowers or seashells" (Félibien 1705: xx). This position was reaffirmed towards the end of the eighteenth century by the English artist Sir Joshua Reynolds (1723–92), who insisted in his *Discourses on Painting and the Fine Arts* that still-life painting did not appeal to the mind's "generalising powers" (Reynolds 1837: 43–4). Throughout the eighteenth century, several writers about art and aesthetics argued against this lowly evaluation (Bryson 1989: 252).

In eighteenth-century France, Jean-Baptiste Du Bos (1670–1742) and Denis Diderot (1713–84), by contrast, promoted an appreciation of still lifes (and, by extension, the representation of plants) as aesthetic objects. Viewers were encouraged to observe the artist's use of color and brushwork in still-life paintings. The aesthetics of imitation became the essential means of understanding such works. Du Bos argued that the subject matter in a still life becomes secondary to the formal qualities of the work: "Tis not so much the object, as the artist's abilities, that draws our curiosity; we bestow no more attention on the object imitated in the picture, than we should on that same object in real nature" (DuBos 1748: 57). The artist's creative application of paint thus makes the viewer rediscover the familiar and appreciate beauty in the mundane. This attitude toward interpreting the representation of plants in still-life painting as aestheticized objects is paralleled in the aesthetic ideas of Immanuel Kant, who would write about the enjoyment he had from observing the colors of a tulip (Marder 2014: ch. 8).

FIGURE 8.5 Jan Davidsz de Heem, *Vase of Flowers*, 1660. Baron Lionel Nathan de Rothschild Collection, National Gallery of Art, Washington, DC.

Perceived as an aesthetic object, a still life representing plants is no longer about the reality outside of the work, be it the references to the Bible or the wealth of the Dutch merchant empire. Instead, it becomes focused on the observer and their ocular perception of the work. The plant allows the artist to explore the way that brushstrokes mix together on the canvas to create an illusion of reality. This visual trick was much admired by Diderot who noted that, up close, the painting represented just a mix of brushstrokes. Diderot emphasized in his writings on the Salon of 1765 that Chardin's art recreated nature through the application of paint: "Chardin's method is very peculiar. It appears to have something in common with something that is thrown together, which when up close one cannot tell what it is, and that as one distances oneself the object is created and becomes nature itself" (Diderot 1995: 303–4). The focus on rendering the colors of plants instead of their contours or other features allowed the artist to create an atmosphere of harmony and restraint in paintings. In the case of Chardin's *Still Life, Flowers in a Vase* (see Figure 8.6), the whites, blues, reds, and pinks in the bouquet are offset by a sandy brown background, enhanced by both the blue and white vase and the bright red petals of the single flower on the ledge. Colors allowed Chardin to create an emotional atmosphere to affect the viewer: he "uses colour ... but paints with feelings" (Roland Michel 1996: 138).

This attention to representing color and light evidenced in the representations of plants is articulated in Isaac Newton's ideas on color in *Optiks* and John Locke's writings

FIGURE 8.6 Jean-Baptiste-Simeon Chardin, *Still Life, Flowers in a Vase*, c. 1760–3. Scottish National Gallery, Edinburgh, UK. © Alamy.

on visual perception in *Essay Concerning Human Understanding*. In Chardin's still life the flowers are distinct, whereas the rest of the work is undetailed, making it a literal representation of an act of seeing which does not treat all the observed objects with the same level of attention (Baxandall 1985: 77). Through his brushwork the artist is communicating to the viewer what he wants them to focus on, in this still life the natural beauty of petals instead of the man-made beauty of the porcelain vase.

Chardin's handling focuses the viewer's gaze on the flowers, which represent the most exciting section of the work, bursting with energetic brushstrokes. He avoided featuring the plants' stems and leaves, and as a consequence the flower petals emerge straight from the vase, making the representation of nature an extension of art. Chardin's vivid representation of plants, in contrast to his rendering of the vase, implies the artist's attitude to nature, which Diderot argued should be studied intensely if one wished to comprehend the purpose of the painting's extraordinary effects: "The eye must be taught to look at nature; and many are those who've never seen it and never will!" (Diderot 1995: 1:4–5). The contrast between the harmonious meditative (and plain) background and the energetic brushstrokes that describe the flowers is a call to study nature with fresh eyes and vigor. Chardin's still life stands as a work about the beauty of nature turned into art by the hand of the artist. By selecting a limited number of flowers representing three colors, placing them in a beautiful vase and obscuring the stalks, the artist turned common flowers into something extraordinary. This process of the aestheticization of plants through representation aligns with the Enlightenment treatment of nature, according to which only nature shaped by art was beautiful and worth contemplating (Delon 2013: 1047–8).

Even though Diderot promoted Chardin's still-life painting as purely aesthetic images, it was still possible for a viewer to interpret its subject matter according to their own more common and local experiences. Philosopher and mathematician, Yves Marie (Père) André (1675–1764) wrote in his *Essay on Beauty* (1741) that "there is a kind of beauty in expression which comes from saving as much as will allow *personnes d'esprit* the pleasure of supplying the rest" (André 2010: 54). Providing visual evidence of this critique, Chardin created still lifes that were more subtle and open to interpretation than were those in the seventeenth century. The plants chosen for this work are rendered in a summary manner which prevents them from being identified as anything other than common garden plants. This differentiates Chardin's work from earlier Dutch still lifes with which his art was often compared (Bryson 1989: 248). In Dutch art, each plant could be easily identified and had its own meaning. Whether valued for its treatment of color or form, truth to nature, or simple depictions of familiar flowers, the subject and its treatment similarly disavow the traditional emblematics, or learned symbolism, associated with the representation of plants.

PLANTS IN PORTRAITURE

While botanical studies and still-life subjects provided artists and viewers with the most focused opportunities to consider the representation of plants, other genres of art regularly included a variety of plants to contribute additional meanings to their primary subjects. Conventions within royal portraiture continued to include plants to signify power (laurel branches) and wisdom (oak), as well as to symbolize national origin (such as the lily in France) and peace brought about through treaties and arranged marriages (olive branches and palms). During the Enlightenment, however, plants took on additional meanings that were more closely tied to the individual, with interpretation directed by other elements surrounding the sitter.

Portraits of rulers, whether of men or women, most commonly drew on inherited forms of traditional emblematics, largely due to the expected formality of the genre. The portrait of *Louis XV giving Peace to Europe* (1729; Chateau of Versailles, France) by François Lemoyne (or Le Moine, 1688–1737), for example, includes plants associated with Roman emperors to reinforce the French king's authority and lineage (Welch 2017). It memorializes the Treaty of Seville (1729) that ended the Anglo-Spanish War. Louis XV, dressed in Roman military clothes, is handing down an olive branch to Europa, personified in female form. The King's role in international politics appears as a type of devoted service to this idealized figure. The olive branch is both a reference to the glories of Imperial Rome and a symbol of the future prosperity of Europe, as the branch bears fruit about to ripen. Next to the King, a nursing woman wears an olive wreath signifying that peace brings nourishment and prosperity.

Despite gender differences, Catherine the Great is frequently represented with the same plant symbols that accompanied her male counterparts. This may be explained by her need to assume masculine modes of behavior to secure her power, authority, and legitimacy on the throne. In the *Portrait of Catherine II the Legislatress in the Temple of the Goddess of Justice* (1783; The State Russian Museum), by Dmitry Levitsky (1735–1822), Catherine is shown pointing towards burning poppies, which symbolize her self-sacrifice of inner peace (*pokoi*) for the benefit of the nation. The plant, opium poppy, was a symbol of sleep and rest going back to the traditions of Ancient Rome, when the god Somnus was depicted with a bunch of poppies (Levitsky 1783: 18; Meiss 1966: 358). An eagle by her side is holding a laurel branch, a symbol of victory (Impelluzo 2004: 38).

Plants can also be seen in many portraits of Maria Theresa of Austria (1717–80), yet they all relate to her own life, requiring simultaneous reference to her authority to rule and capacity as a woman to bear children. The portrait *Archduchess Maria Theresa* by the Danish painter, Andreas Möller (1684–*c.* 1762), depicts the future empress with flowers to symbolize both her association with Austria and her femininity (1727; see Figure 8.7). White and orange flowers sit in the folds of her dress, referencing her fertility, an essential purpose for a woman of royal origin. The most prominent flower in her skirt is a rose, a symbol of Virgin Mary, referring to the sitter's piety and purity (Yonan 2011: 23, 193). The plants also serve to highlight the beauty of her pale skin and gentle facial features, while the manner in which she holds the narcissus displays her grace. She is shown taking the flowers from her skirt and placing them in the bouquet underneath the crown signifying the royal destiny of her future offspring. As the flowers in the bouquet bloom at different times of the year, the fact that they are all flowering simultaneously signifies the power of royalty to influence nature and make it suit their needs.

Less stifled by the need to adhere to the conventions of court portraiture, the representation of plants within informal portraits of the nobility took on new connotations in step with major currents of social change during the eighteenth century. Philosophical writers established a connection between plants, charity, and education by drawing useful analogies through which to express ideas related to the "nature vs. nurture" debates over a child's development. In his pedagogical treatise *Emile* (1762, para. 13, book 1), Rousseau asserted that "Plants were fashioned by cultivation and men by education." Similar ideas are expressed visually in a portrait of Russian nobleman Prokofiy Demidov (1710–86) by Levitsky (1773; see Figure 8.8). The painting was modeled as a mock portrait to poke fun at the tradition of Grand Tour portraiture, specifically the portrait (1766) of Count

FIGURE 8.7 Andreas Möller, *Archduchess Maria Theresa*, 1727. Kunsthistorisches Museum, Vienna. © Alamy.

Kirill Razumovsky by Pompeo Batoni, in which the sitter points towards iconic sculptures he saw during his journey (Antonova 2005). Demidov adopts the pose of his compatriot, but wears casual bedtime clothes in contrast to the sumptuous suit worn by Razumovsky. More importantly for understanding the representations of plants within portraiture of this period, instead of gesturing to classical sculpture, the sitter points towards two pots with flowers situated next to the colonnade.

FIGURE 8.8 Dmitry Levitsky, *Portrait of Prokofiy Demidov*, 1773. State Tretyakov Gallery, Moscow. © Alamy.

Although the painting realizes the comic effect of mockery, there are several other meanings to be gleaned from the portrait and the choice and placement of plants in the composition. At a superficial level, the potted plants signify Demidov's passion for botany. As one of the wealthiest men in Russia, he created an exceptional private botanical garden near Moscow and spent much of his time botanizing and creating herbarium specimens. Yet with more reflection on the sitter's life, it becomes clear that the plants hold deeper meanings, that refer to Demidov's charitable activities. In the background there is a foundling home, a charitable educational establishment that he supported. The portrait is an allegory of a cultivated educational upbringing he wished to create for the students there. In this context, an open book refers to the benefits of an enlightened interest in reading, while his informal clothing indicates that this is his personal hobby. The image overall shows a progression from a bulb (abandoned child) combined with a watering can (care) and a botanical book (knowledge), on the sitter's left, to the plants in pots on his right, one being the same as that figured in the botanical book, a polyanthus, the other a thornless rose. The polyanthus is a hybrid (*Primula* x *polyantha*) with the wild European primrose (*P. vulgaris*) in its ancestry. The didactic message that this plant perhaps communicates to the viewer is that care and education could turn an orphan (wildflower) into a cultivated individual. The rose has buds about to open, referring to the potential of education to nurture the growth of the child, especially when viewed in tandem with the plant's thornless representation here. Combined with two columns, these plants create a message of strengthening the state through the education of future generations. This sentiment was popular in the eighteenth-century Russia, where tsars Peter I and Catherine II were frequently depicted and described as gardeners planting the "gardens of sciences" through their reforms, which would help Russia to prosper (Baehr 1991: 80–2).

The rise in voyages of discovery during the eighteenth century and the resulting imperialist plans of many European countries influenced botanical displays within artworks and introduced a new type of portraiture involving plants: the memorialization of the sitter as adventurer. Both Levitsky's portrait and that of American artist Benjamin West (1738–1820) depicting Joseph Banks (1773; Usher Art Gallery, Lincoln, UK) manipulate the tropes of Grand Tour portraiture. As Demidov had gone on a Grand Tour, Levitsky playfully deployed this fact from his sitter's biography in the pose and format of the portrait to suggest that the practical and metaphorical cultivation of plants by a nobleman could benefit one's cultural sophistication, as much, if not more than travel.

West's portrait of Banks engages with the conventions of Grand Tour portraiture on different terms. It represents Banks' travel as an explorer and botanical collector in superior terms to that of mere Grand Tourists, because his tour of the globe allowed him to see more of the world, complete with the pleasures and dangers of unknown discovery, in contrast to the relatively safety of following the established routes of a European Grand Tour (Coleman 2005: 13). Rather than showing typical classical sculptures purchased during such Tours, Banks chose to display exotic items that would have been instantly recognized as being outside the European experience. Both portraits contain botanical illustration; however, in the painting by West, its meaning changes from charity to colonial botany, a practice of seeking, gathering, and transplanting plants, in some cases those useful for the imperial economy, besides collecting, importing, and growing exotics for their own sake. As a passionate botanist, colonial botany was a practice publicly championed by Banks after his return from the *Endeavour* voyage. The painting stakes a claim for the importance of plant collecting to his self-image and self-narrative (Gooding 2017: 7).

The plant featured in a book at his feet is *Phormium tenax*, New Zealand flax, which Banks hoped would enable Britain to become independent from its existing reliance on Russian flax. This plant, which is the subject of the botanical sketch by Parkinson discussed above, was one of the most commercially valuable botanical "discoveries" of the *Endeavour* voyage. Yet, in contrast to Parkinson's representation of the plant in an aesthetic void, the portrait provides the plant with an economic narrative supplied by the sitter's gesture. Banks is pointing to the New Zealand cloak made from the local flax. As with the portrait of Demidov, the artist has set up a visual relationship between the plant and the sitter to indicate the finished product of the plant's development—in this case, a fine and strong cloak made from fibers perfect for the sails of the British Navy and part of Banks' understanding of the political economy of Britain's imperial ambitions involving botanical exploitation. By demonstrating the commercial rewards of voyages of discovery, Banks was promoting the national benefits of scientific research. In this painting, references to plants elevate the sitter from a foppish Grand Tourist to an agent of colonialism, seeking to support the state in every possible way (Fara 2017: 20).

Plants featured prominently in portraits of botanists and botanical artists, visually conveying their interests and beliefs. In Martin Hoffmann's *Linnaeus in his Lapland Dress* (1737; see Figure 8.9), the most influential botanist of the eighteenth century is depicted holding a delicate twinflower or *linnaea borealis* (McPhail 1979: 54). Linnaeus was undoubtedly proud of this plant, which features in most of his portraits, as it marked what he regarded as the pivotal experience of his career—his research voyage to Lapland in 1732 (Thiem 2017: 57). This trip was crucial to Linnaeus' botanical legacy because it afforded the first opportunity to put his new nomenclature and classification system into practice. Linnaeus chose the *linnaea borealis* as his personal symbol when he was ennobled in 1761 (Scott 2010: 361). A portrait by Alexander Roslin (Nationalmuseum, Stockholm, 1775) depicts Linneaus combining his Order of the Polar Star with the twinflower into one hybrid decoration, signifying the evolution of his status from his modest origins to knighthood.

In Hoffman's portrait, the twinflower serves as a reference to the ties of masculine friendship between botanists. A tag is attached to the plant, stating its name as *Linnaea Gronovius*. Jan Gronovius (1686–1762), a fellow botanist and keen supporter of the classification system developed by Linneaus, named *Linnea* in his friend's honor (Thiem 2017: 57). Gronovius commissioned the portrait in the same year that Linnaeus described the eponymous plant in his publication *Critica Botanica* (1737): "named by the celebrated Gronovius ... a plant of Lapland, lowly, insignificant, disregarded, flowering but for a brief space ... named for Linnaeus who resembles it" (Nickelsen 2006: 139). These self-denigrating remarks were undoubtedly meant as a joke between friends. Similarly, the tag refers to the novelty introduced by Linnaeus into the study of botany—binomial nomenclature (the name *Linnaea borealis* was coined by Linnaeus himself). This nomenclatural system simplified and unified the naming of plants by replacing conventional names with a combination of generic and specific names in Latin. Linnaeus renamed a large number of genera after professional European botanists. Binomial nomenclature, which also became standard in botanical illustrations, effectively erased the histories of plants by overwriting their local names (Schiebinger 2004b: 202). By largely excluding women and non-European people from his naming conventions, Linnaeus turned botany into a science of "great" European men. In Hoffman's portrait of Linnaeus, the play on words in the title given to the plant in the image, "*Linnea Gronovius*," served as a reference to masculine friendship, which

FIGURE 8.9 Martin Hoffmann, *Carolus Linnaeus in Lappish dress*. © Photograph by Mikael Wallerstedt, courtesy of The Swedish Linnaeus Society, 1735.

lay at the core of the new nomenclature system. It allowed Linnaeus and other botanists to name pleasant plants after the people they respected and admired, and foul-smelling or unattractive ones after their enemies (Schiebinger 1993: 392). The inclusion of a set of books to the left of the sitter is also telling. These are the volumes Gronovius helped Linnaeus to edit and publish between 1735 and 1737, pictured here as the fruits of their friendship (McPhail 1979: 53). Neatly stacked and positioned next to the twinflower, the publications on botanical classification highlight Linnaeus' attitude towards nature, which combined admiration with objectivity and a desire to bring order to the natural world.

Plants frequently highlight the sitter's origin or location within portraiture. In Joshua Reynolds' *Portrait of Omai* (1776; see Figure 8.10), the sitter was an active participant in the creation of his painted image. This portrait is the result of cultural negotiation between Reynolds and Omai, both of whom saw this commission as a something mutually advantageous. As a portrait of a celebrated Polynesian visitor to England, it was a deliberate exercise in self-fashioning aimed at promoting Omai's distinguished status to garner British support for retrieving his lands from Boraborans once back in the Society Islands (now part of French Polynesia). His right hand is pointing toward the leafy palm signifying his exotic origin and "otherness" within London society (Fullagar 2010; Kelly 2011). Reynolds adopted Omai's gesture from the sculpture of antiquity, once again following the conventions of Grand Tour portraiture. Yet the effect of this visual reference is an inversion of the model. This is the portrait of a tourist who chose not to promote his experience of travel or the knowledge he gained during it. Instead, Omai is looking back at his origins in Tahiti and prominently displaying his exotic status by showing his tattoos, tapa cloth robes, and choosing to be depicted in a location that recalls the distinctive landscape of his native lands (as opposed to the familiar landscape of Europe).

The palm in the portrait is rendered as a generic concept, without specificity, using broad brushstrokes. The botanical identification is not only impossible, but, more importantly, irrelevant and a potential distraction from the higher ideals of the painting, as Reynolds argued in his *Discourses*, a series of lectures delivered at the Royal Academy between 1769 and 1790. Assisted by the theoretical aims of Reynold's elevated approach to portraiture, the portrayal of Omai effectively reversed the extractive policies of Empire, which removed plants and people from their native environments and erased their cultural differences. Through the inclusion of exotic plants in the portrait of an "exotic" sitter, Omai is shown demonstrating his independence from the imperial venture and taking pride in his culture. With this image, the sitter inverts the cultural hegemony of European culture and imperial enterprise asserting his independence and self-respect.

In contrast to those in portraits of men, images of plants in portraits of women are characterized by their physiological characteristics, such as fertility, virginity, or beauty. Male portraits from the period tend to have the entire plant displayed, whereas those of females often include cut flowers, which consequently become disassociated from the plant itself. Flowers held or strewn across the lap suggested fertility and loss of virginity, at the same time referring to the sitter's beauty. Eighteenth-century portraits continued the long-established tradition of equating women with flowers, both physiologically and psychologically. Yet it was this very association that enabled botanical science and art to be perceived as an ideal pursuit for women, advocated by philosophers like Rousseau.

Being pictured in a garden or pastoral setting acquired a new connotation in the second half of the seventeenth century. The pastoral mode allowed noble women to assume an identity different from their subservient role at court. Portraits in a pastoral guise allowed

FIGURE 8.10 Joshua Reynolds, *Portrait of Omai*, 1776. Oil on canvas. Collection of John Magnier. © Sothebys.

female sitters to highlight their cultural and intellectual accomplishments, as opposed to more generic celebrations of their marriageability and beauty. The popularity of garden retreats also contributed to this mode of portrayal. Health benefits were attributed to these spaces by the neo-Hippocratic medical revival, which emphasized the ameliorative effects of the rural environment on mental and physical health (Martin 2011: 118). These concerns purportedly influenced Madame de Pompadour to commission a portrait from Charles-André van Loo (1705–65), depicting her as a gardener (*c.* 1754–5; Chateau of Versailles, France).

Pompadour's portrait references the special favor she enjoyed as mistress of the French king Louis XV, who gave her a Hermitage at Versailles. In her right hand she grasps jasmine, while also holding a basket of carnations, roses, and buttercups. The artist balanced the color scheme of the portrait to align her skin with flowers, her lips with the roses, and ribbons with the sky, suggesting that she was living in harmony with nature. In this instance, flowers work to articulate a message of the sitter's health and influence, aimed to counter the gossipy insinuations of her enemies at court. The choice of white jasmine was a deliberate attempt to demonstrate the beauty of Pompadour's skin, which at that stage was widely criticized for being yellow. Moreover, it was an attempt to address accusations that she had a venereal disease called white flower (Burrows 2010: 108; Martin 2011: 156). By holding a white flower in her hand, she was refuting such accusations through a reference to her glowing complexion and wholesome leisure pursuits.

ROCAILLE DECORATION AND PLANTS IN ARCHITECTURE

While plants adorned portraits, often with purposeful meanings, and assumed a primary place of purpose as the subjects of still lifes and botanical images, they also became the foundational forms of the most innovative development in architectural decoration of the period historically known as the Rococo (Milam 2011). Derived from the term *rocaille* (or rockwork), the style incorporated forms modeled on nature but transformed through curvilinear ornamentation (Kimball 1980). *Rocaille* decoration freed itself from the traditions of the past and pointed toward more modern forms of architectural design, devoid of structural purpose. At once unprecedented and historically informed, these novel Rococo interiors idealized a fictional pastoral life of nobility's past, implying a contrast with the formality of court in the present. Walls were decorated with verdant arabesques, protectively wrapped around shepherds and shepherdesses, suggesting faithfulness and fertility in its very lushness of ornamental form.

Some patrons went further in their pursuit of the pastoral ideal. Their interiors emphasized the regeneration of the natural and social order secured through the visualized superiority of the nobility. In these interiors, the inhabitants' ability to make sense of the confusing elements of architecture enacted a rarefied aesthetic understanding, which would allow them to lay claim to power and influence. This effect was achieved through the use of *trompe l'oeil* murals with garden views proposed by designers like Gilles-Marie Oppenord[t] (1672–1742) (see Figure 8.11). These murals removed the barriers between image and reality, form and function, implying the noble owners' unique capability to distinguish falseness from truth. Rococo artists occluded vanishing points to surprise and disorient beholders. Experienced visitors to these spaces would have felt a pleasant sensation after bringing unity to the disorderly world around them.

FIGURE 8.11 Gabriel Huquer after Gilles-Marie Oppenord, designs for salons, *c*. 1748. Etchings. © Bibliothèque nationale de France, Paris.

In tandem with these *rocaille* interiors, eighteenth-century designers and architects introduced more naturalistic decorative elements derived from plants, stimulated by the wide-spread interest in collecting and displaying natural history specimens. Enlightenment ideas of empiricism encouraged patrons to perceive material reality as knowable through observation, a skill in which noble amateurs considered themselves to be superior to the growing class of learned professionals. The nobility adapted the fashionable pursuit of natural sciences as a way of modernizing the pastoral ideal. In the new interiors, palms and garlands of flowers—the most popular decorative elements—were used literally to tie together displays of arts and sciences. Through combining the disparate items, these spaces promoted a new system of knowledge that united pastoralism and natural history, one promoted as similar to that enjoyed by Adam in Eden (Scott 1995: 161–75).

Exotic plants, like pineapples and palms, became popular elements of garden architecture. This trend was influenced by the growth of European empires and the spread of ideas of cosmopolitanism (Milam 2012). Garden owners were encouraged to travel in their mind by visiting different pavilions. Plants were particularly prominent in the decoration of *chinoiserie* garden structures. The *Chinese Teahouse* (1755–65) in Potsdam by Johann Gottfried Büring (1723–88) was adorned by twelve gilded sandstone columns in the shape of palms (see Figure 8.12).[4] In a fantastic stroke of artistry, these plants transferred the Chinese landscape into a southern climate.[5] The pavilion's façade

FIGURE 8.12 Johann Gottfried Büring, Chinese House, Sanssouci, Potsdam, Germany. Wikimedia Commons. Photo by Johann H. Addicks.

was also decorated with pineapples, considered to be a symbol of luxury and exotic origins (O'Connor 2013; Walvin 1997; Welch 2017; and Chapter 2 of this volume).[6]

Such a playful engagement with the idea of empire was further developed in the Dunmore pavilion, constructed in 1761 for John Murray, 4th Earl of Dunmore in Stirlingshire, Scotland. The wings of this building were occupied by pineapple hothouses with the central section, situated underneath the pineapple, serving as a pleasure folly (Gohmann 2018: 154–5). Dunmore's guests were invited to look out from this level upon the hothouses with peaches and other exotic fruits growing in the garden. This architectural scheme emphasized the centrality of pineapple hothouses which protected the most rare and challenging fruit crops produced at the time. The combination of the neoclassical bottom section with a pineapple dome symbolized the desire to "naturalize colonial transplantation" by literally merging the colonies and the metropole in the single building (Casid 2005: 53–4; Gohmann 2018: 143–61). According to Jill Casid, this hybridization of both gardens and architecture demonstrated a perceived "natural" relationship between the colonies and the motherland, helping to promote Britain as a colonial power.

Like other forms of representation during the seventeenth and eighteenth centuries, architectural ornamentation served to attach a multitude of meanings to plants that were understood on a variety of levels by those who viewed and experienced these spaces at the time: exoticism combined with domestic purpose and use; imperialism promoted botanical interest and an expansion of knowledge, while the development of aesthetics in philosophical circles urged the visual engagement of plants as a source of pleasure and intellectual contemplation. While all of these impulses underscore the changing nature of plant representation in the period, it is the latter which exerted the most influence over the directions of art-making in the centuries that followed.

NOTES

Chapter 1

1 In old English currency, "d" = pennies and "s" = shillings; one shilling is equivalent to twelve pennies; one penny divides into four farthings or two halfpennies.

2 Calculated according to Assize tables with the assistance of William Rubel; see Penkethman (1638) and Penkethman and Whitaker (1648).

3 Devonshiring is the process whereby the top layer of turf is skimmed off using a breast plow, the turf piled into small heaps which are set alight and left to smoulder, and the resulting ash ploughed in before manuring.

4 It is difficult to calculate equivalent monetary values over such a long time period. The range suggested here is based on the Measuring Worth calculator at https://www.measuringworth.com/calculators/ukcompare/ (the lower end) and the Bank of England inflation calculator at https://www.bankofengland.co.uk/monetary-policy/inflation/inflation-calculator (the upper end).

Chapter 2

1 Emperor Qianlong's Letter to King George III, 1793. Reproduced in the Modern History Sourcebook at http://www.fordham.edu/halsall/mod/1793qianlong.html</URI> (accessed May 24, 2021).

2 It has also been proposed that the first pineapple in Europe was cultivated at Meerburg in the 1650s by Pieter de la Court (1618–85), father of the younger Pieter de la Court discussed below in relation to hothouses. Modern historians treat this as a "legend" as there is no evidence that this was the case. It may be that there was some confusion over the pursuits of the two de la Courts, with the son's interests more squarely focused on gardening. See Weststeijn (2012: 284).

3 Grigoriy Pilnikov, *Plan of the peach and pineapple hothouse*, around 1787, Pavlovsk Museum, Inv. no. Ch-569. My thanks to Dr. Ekaterina Heath for sharing her archival research on Russian gardens.

Chapter 3

1 The 1726 proclamation has been mentioned by several historians studying eighteenth-century French botanical collecting, but with little attention to the limited extent to which its instructions were actually applied. For more on this, start with Mukerji (2005: 27) and Paskvan (1971: 86–90).

2 As Hahn makes clear, at certain times in its history the Académie was made to answer to the aspirations of certain royal ministers rather than the King directly. Colbert's successor Louvois, for example, aligned the scientific progress of the Académie with his own projects.

3 McClellan and Regourd discuss some of the ways in which the state was circumvented in "The Colonial Machine: French Science and Colonization in the Ancien Régime" (2001: 44–8). They conclude the section, however, by lessening the importance of these "secondary centers and modes of activity," and reaffirming the centrality of "the great and powerful French colonial machine" (48), an interpretation which I do not fully agree with.

4 McClellan and Regourd concluded that French merchants showed little interest in studying the natural world (2001: 48–9). However, more recent research into the activities of the various European East India Companies suggests that this assertion should be reassessed. See contributions to Damodaran et al. (2015) and Winterbottom (2016).

5 The correspondence between André Thouin and Jean Nicolas Céré clearly shows Thouin's acknowledgment of Céré's expertise. The letters in the Muséum National d'Histoire Naturelle, Paris—MS307, for example—demonstrate Thouin's dependence on Céré for guidance and information about the climate in India, even though, of course, the vast Indian subcontinent was still several weeks away by sea.

6 As a further example, E. C. Spary (2000: ch. 2) has underlined the significance of the patronage system—and thus of private funding and the value of honorific and symbolic rewards—to Enlightenment botany.

7 For more on plant collectors and taxonomy, see Chapter 4 of this volume by Alexandra Cook.

Chapter 4

1 The "East Indies" referred to modern-day India, Sri Lanka, Indonesia, the Philippines, and the Marianas; China, Japan, and Vietnam might also be covered by this term. The "West Indies" referred to the Americas and the "Indies" to all of the above (Cook 2010: 124).

2 "Subaltern" refers to "native non-white and other colonial subjects" (Eamon 2018: 145).

3 Herbals, i.e. textual compendia of plant knowledge, especially medicinal, should not be confused with herbaria, which are dried plant collections.

4 These figures are only indicative, because there was no settled species concept (Wilkins 2011).

5 However, folk names do not necessarily reference medical uses since medicine in many cultures constitutes closely guarded secret knowledge (Cook 2010: 129; Kathirathamby-Wells 2015; Mahr 1955: 11–12; cf. Schiebinger 2004b: 196, 220–2).

6 It has to be acknowledged that "the West's development of a worldwide biological systematics explicitly involved disregard of ecological relationships and of the colors, smells, sounds, tastes and textures" on which folk classifications rest (Atran 1999: 181).

7 Commemorated by the genus, *Rauwolfia*, Rauwolf was the first European to describe coffee preparation and public baths of the Near East (Dannenfeldt 1968). His account of this journey went through three editions (1582–3) (Rauwolf 1582).

8 "Aureliana" refers to Philippe d'Orléans (1674–1723), Regent of France during the minority of Louis XV, to whom Lafitau dedicated his account of North American ginseng (*Panax quinquefolius*) (Lafitau 1718).

9 *Pan Suecicus* was a doctoral dissertation, written, as was usual, by the professor and only defended by the candidate (Stearn 1957: 51–5).

10 Antoine-Laurent de Jussieu is referring to the nomenclatural reform he carried out in the Jardin du Roi (Paris), now Jardin de plantes, in 1774.

11 A survey of the etymologies of all names in *Species* is beyond the scope of this chapter.

12 Elderly women who supervise the selection of rice for planting distinguish among over five hundred cultivars, each of which is grown under unique conditions on terraced mountainsides. Personal communication, Gerald W. Sullivan.

13 For example [Leonhart] Fuchs (1501–66), Otto Brunfels (1488–1534), Hieronymus Tragus (1498–1554), Carolus Clusius (Charles de l'Ecluse) (1526–1609), Mathias de L'Obel (Lobelius) (1538–1616), and Rembertus Dodoens (1517–85).

14 The ancients include Dioscorides (*c.* 40–90 CE) and Galen (130–210 CE), together with the encyclopaedist, Pliny the Elder (23–79 CE). Dioscorides' *De materia medica* documents 520 plants, of which fifty-four are considered essential medicinal plants by the World Health Organization.

15 Theophrastus' was also a folk classification, but one transmitted to Europe from Greek antiquity.

16 "Malabar" then referred to the entire Tamil region of southern India (Neumann 2010: 201n13).

17 Recent accounts (e.g. Raj 2017) ignore Reede's use of the *Hortus* in an internal VOC dispute regarding transfer of power from Malabar to Ceylon (present-day Sri Lanka).

18 Grove (1996) does not substantiate his oft-cited claim that southwest Indian Ezhava classification features in *Hortus Malabaricus* (Raj 2017). In fact, Ezhava classification played a minor role because the work was assembled by Dutchmen influenced by Bauhin (Heniger 1986). Singh (2015: 194) claims the local classification is unknown because Reede's notes are lost.

19 Hermaphrodism already interested the ancients, as evidenced by Plato's *Symposium*.

20 Strained analogies to the animal kingdom pushed sexualism to the limit (Delaporte 1982: 127–36).

21 Similarly, the nectary was long considered merely excretory (Lorch 1978: 529–32).

22 This work exemplifies the vernacularization of scientific texts in this era, with the Latin version following the French. Valeriani and Ishizu (2015) discuss Chinese as an erudite language in Japan.

23 Antoine de Jussieu was the older brother of the founder of the natural family system of classification, Bernard de Jussieu (1699–1777). Both were uncles to Antoine-Laurent de Jussieu, who expanded on his uncle Bernard's work in developing a natural method of classification. Antoine de Jussieu sponsored his younger brothers' medical studies, thereby launching the botanical dynasty that played a significant role in modern taxonomy and dominated the Jardin du Roi (Jardin des plantes) into the nineteenth century.

24 Vaillant claimed the work, published in United Provinces of the Netherlands, appeared without his permission (Guédès 1983: 195–7).

25 Cf. Ereshefsky (2001: 202) who inaccurately claims Linnaeus arrived at 3,884 combinations based on thirty-one parts.

26 This distinction was first broached by the ancient atomists and elaborated by Early Modern philosophers such as Locke and Berkeley (1997: 135ff.).

27 A notable exception was Adanson, who devized his own natural classification and nomenclature (Adanson 1966).

28 Koerner (1999) highlights Linnaeus' cameralism: "an extreme form of mercantilism, concentrating ... on building up state power, and subordinating all parts of the economy and polity to the state and its bureaucracy," https://mises.org/library/who-were-cameralists (accessed May 23, 2021).

29 For example, "the fundamental divisions Linnaeus devised as the basis of his new botanical taxonomy recapitulated the sexual hierarchy of Western Europe" (Schiebinger 1991: 123).

30 The first three volumes of Rauwolf's herbarium contain European plants, each volume comprising 106 leaves, with plants glued on each side, framed by cardboard to prevent specimens from touching. The fourth volume holds 364 Levantine specimens, glued to one side of the page, with plant names in Latin and/or German, Arabic, and local dialects. Rauwolf's herbarium was looted from Munich by Swedish troops during the Thirty Years War, and given by Sweden's Queen Christina to her librarian, Isaac Vossius; in 1688 the herbarium was purchased by Leiden University. A list of its contents ordered according to Linnaeus' artificial sexual system was published in 1755 (Gronovius 1755).

31 I owe this reference to Professor Abdesselam Cheddadi, Rabat, Morocco.

32 Harder's herbarium, which entered the collections of the Grand Dukes of Bavaria, is now in the Staatsbibliothek Munich. Available online: https://bildsuche.digitale-sammlungen.de/index.html?c=band_segmente&bandnummer=bsb00011834&pimage=00022&l=de (accessed May 23, 2021). Rozier's is in the Municipal Library, Lyon, but only partially digitized.

33 Darwin's studies in comparative morphology, essential to his later evolutionary theory, depended on specimen collections as essential complements to field work (Winsor 2009: 49).

34 The natural history collections were later moved to a purpose-built British Museum (Natural History), now the Natural History Museum in London.

35 Murphy (2013) points out the interconnections between the slave trade and Petiver's collections. Note that Kamel was not a priest, even though he is often described as "Father" Kamel.

36 Alkaloids include caffeine, cocaine, strychnine, quinine, nicotine, and morphine, among others.

37 Plants hold huge potential in treating malaria, antimicrobial-resistant superbugs, lifestyle diseases, and cancers. Successes include anti-cancer agent, Pacific yew, *Taxus brevifolia*, and anti-malarial artemisinin (from *Artemisia annua*), discovered in ancient Chinese texts by Nobel prize winner, Tu You You.

38 The word "scientist" was not in use before the nineteenth century.

39 See Chapter 2 of this volume.

40 The breadfruit, thrown overboard by the *Bounty* mutineers, was intended as food for slaves in the British Caribbean. Bligh's second voyage in command of HMS *Providence* (1791–3) successfully executed this intra-tropical plant transfer.

41 Unlike the eighteenth-century terms "Enlightenment," "lumières," or "Aufklärung," "scientific revolution" first arises in the 1940s (Eamon 2018: 142).

42 "Agnotology" is the study of knowledge erasure via culturally constructed ignorance (Proctor 1995; Schiebinger 2004b, 2017). Ironically, subaltern contributions to scientific naming have suffered erasure at the hands of those keen to combat agnotology.

Chapter 7

1 For the letters of Jemima, Marchioness Grey, see Twigs Way, unpublished "Report" for English Heritage, letter to Catherine Talbot June 2, 1748: access on request through English Heritage, London, www.english-heritage.org.uk.

2 Linnaeus created a plant taxonomy based solely on the number and arrangement of the reproductive organs, which resulted in many groupings that seemed unnatural. He freely admitted that this produced an "artificial classification," not a natural one. Later systems of classification would return to John Ray's use of morphological evidence from all parts of an organism in all its stages of development—a "natural classification."

Chapter 8

1 Aquatint and Mezzotint are both intaglio techniques by which an image is incised or etched into a metal plate. Ink is applied to the plate and fills the sunken areas before printing onto paper. Traditional engravings use a series of hatched (parallel) or cross-hatched lines to create shades. Aquatints, in contrast, produce tonal variations by using acid to eat into the printing plate, creating submerged sections that hold the ink. Mezzotints use a tool called a rocker to fully abrade the metal plate before printing, which tends to blurr the precision of lines. Tonal variations are created by scraping and burnishing areas of the plate, so that lighter tones emerge from an overall darker image.

2 Museum of New Zealand Te Papa Tongarewa Herbarium, Cook's First Voyage Collection, Inv. 63573.

3 Banks planned to publish these botanical illustrations in his *Florilegium* documenting the flora of the continents he visited during the voyage. Due to the high volume of plants that needed to be drawn at the same time, Parkinson developed a method of outlining the plant and capturing the colors of its key components. His premature death prevented him from finishing the watercolors, which were later completed by artists whom Banks hired. In 1789, Frederick Nodder completed *Phormium tenax*, according to Parkinson's original color markings (Lack 1997: 88–90).

4 Palms also decorate façades of the Carl Fredrik Adelcrantz's *Chinese pavilion*, Drottningholm, Sweden, 1769.

5 In 1789 architect Heinrich Ludwig Manger wrote that "the house would not be considered characteristic and decisive enough as an Asiatic design if the palm trees were not here to suggest the climate" (Manger, *Baugeschichte* vol. I, 237, trans. in Jacobson 1993: 94).

6 A pineapple is featured on the roof of one of Chinese pavilions in Le Rouge's *Détail des nouveaux jardins à la mode*; pineapples also decorate the Chinese pavilions in Lustgarten, Weitsheheim (1763–74).

BIBLIOGRAPHY

Adanson, Michel. [1763–4] 1966. *Familles des plantes*, 2 vols. in one, reprint edn. with an introduction by Frans A. Stafleu. Lehre: J. Cramer.

Agar, Ben. 1641. *Clericus mercati, &c.* London.

Aiton, William. 1789. *Hortus Kewensis*. London.

Albala, Ken. 2003. *Food in Early Modern Europe*. Westport, CT: Greenwood Press.

Albertone, Manuela. 2009. "Thomas Jefferson and French Economic Thought: A Mutual Exchange of Ideas." In M. Albertone et al. (eds.), *Rethinking the Atlantic World. Europe and America in the Age of Democratic Revolutions*, 123–46. London: Palgrave Macmillan.

Albritton Jonsson, Fredrik. 2013. *Enlightenment's Frontier. The Scottish Highlands and the Origins of Environmentalism*. New Haven and London: Yale University Press.

Alpers, Paul. 1982. "What is Pastoral?" *Critical Inquiry*, 8(3): 437–60.

Alpers, Paul. 1996. *What is Pastoral*? Chicago: Chicago University Press.

Alyon, Pierre Philippe. 1787–8. *Cours de botanique, Pour servir à l'education des enfans de S.A. Sérénissime Monseigneur Le Duc D'Orléans, où l'on à rassemblé les Plantes Indigènes et Exotiques employées dans les Arts et dans la Médecine*. Paris: Chez l'auteur.

Anagnostou, Sabine. 2007. "Jesuits in Spanish America: Contributions to the exploration of the American materia medica." *Pharmacy in History*, 47(1): 3–1.

Anderson, E.N. 2014. "China." In Paul Freedman, Joyce E. Chaplin, and Ken Albala (eds.), *Food in Time and Place: The American Historical Association Companion to Food History*, 41–67. Oakland, CA: University of California Press.

Anderson, Frank J. 1997. *An Illustrated History of the Herbals*. New York: Columbia University Press.

André, Yves-Marie. 2010. *Essay on Beauty*. Translated by Alan J. Cain. Porto: University of Porto.

Anishanslin, Zara. 2016. *Portrait of a Woman in Silk: Hidden Histories of the British Atlantic World*. New Haven and London: Yale University Press.

Anon. 1642. *An Abstract of Severall Letters from Hull, York, and Beverly, of his Majesties proceedings*. London: For Ben. Allen.

Anon. n.d. [*c.* 1640]. *A new ballad, shewing the great misery sustained by a poore man in Essex*. London: For H. Gosson.

Anstey, Peter R., and Stephen A. Harris. 2006. "Locke and Botany." *Studies in History and Philosophy of Science Part C: Studies in History and Philosophy of Biological and Biomedical Sciences*, 37(2): 151–71.

Antonova, Irina. 2005. *Rossiya- Italia. Skvoz' veka. Ot Giotto do Malevicha*. Moscow: Khudozhnik i kniga.

Arber, Agnes. 1986. *Herbals: Their Origin and Evolution, A Chapter in the History of Botany, 1470–1670*, 3rd edn. Cambridge: Cambridge University Press.

Archives départementales de Loire-Atlantique (ADLA) C632, Folder #2: Lettres patentes du Roy, Marly le 9 mai 1708; Ordonnance du Roy, Fontainebleau, 20 septembre 1726; Bonamy, Francis. 1765. [Untitled pamphlet addressed to the États de Bretagne].

Nantes: Imprimerie d'André Querro; Note joint à la lettre de M. de Beaupeau du 16 avril 1786; Drogues nécessaires pour la composition des coffres de medicaments des Bâtiments du Commerce destinés pour les voyages de long cours. n.d.

Archives municipales de Nantes (AMN):DD50, Jardin des plantes de Nantes; DD51, Jardin des apothicaires.

Army of Scotland. 1640. *Our Demands of the English Lords Manifested.* Amsterdam: Printed by Margery Mar-Prelat.

Arredondo, Jaime Marroquín. 2019. "The Method of Francisco Hernández: Early Modern Science and the Translation of Mesoamerica's Natural History." In Jaime Marroquín Arredondo and Ralph Bauer, *Translating Nature: Cross-Cultural Histories of Early Modern Science*, 45–69. Philadelphia: University of Pennsylvania Press.

Åsberg, Marie, and William Stearn. 1973. "Linnaeus's Öland and Gotland Journey 1741." *Biological Journal of the Linnean Society*, 5(1): 1–107.

Aslanian, Sebouh. 2011. *From the Indian Ocean to the Mediterranean: The Global Trade Networks of Armenian Merchants from New Julfa*, vol. 17. Berkeley and Los Angeles: University of California Press.

Atasoy, Nurhan. 2007. "Links between the Ottoman and the Western World on Floriculture and Gardening." In Michel Conan and W. John Kress (eds.), *Botanical Progress, Horticultural Innovation and Cultural Changes.* Washington, DC: Dumbarton Oaks Research Library and Collection.

Atran, Scott. 1990. *Cognitive Foundations of Natural History: Towards an Anthropology of Science.* Cambridge University Press.

Atran, Scott. 1999. "Itzaj Maya Folkbiological Taxonomy: Cognitive Universals and Cultural Particulars." In Douglas L. Medin and Scott Atran (eds.), *Folkbiology*, 119–203. Cambridge, MA: MIT Press.

Aublet, Jean-Baptiste Christophe. 1775. *Histoire des plantes de la Guiane françoise, rangées suivant la méthode sexuelle.* London and Paris: Pierre-François Didot jeune.

Austen, Jane. 1997. Letter to Cassandra Austen, Sunday 8th–Monday 9th February, 1807, in Deidre Le Fay (ed.), *Jane Austen's Letters*, New Edition. Oxford: Oxford University Press

Bacon, Francis. 1638. *History Naturall and Experimentall.* London: John Haviland.

Baehr, Stephen. 1991. *The Paradise Myth in Eighteenth-Century Russia: Utopian Patterns in Early Secular Russian Literature and Culture.* Stanford: Stanford University Press.

Bailey, W.W. 1875. "About Weeds." *The Aldine*, 7(14): 270–1.

Balic, Adam. 2013. "The Haggis." In Mark McWilliams (ed.), *Proceedings of the Oxford Symposium on Food & Cookery 2012: Wrapped and Stuffed Foods*, 82–93. Totnes, Devon: Prospect Books.

Banks, Kenneth J. 2003. *Chasing Empire Across the Sea: Communications and the State in the French Atlantic, 1713–1763.* Montreal: McGill University Press.

Banks, Kenneth J. 2005. "Communications and 'Imperial Overstretch': Lessons from the Eighteenth-Century French Atlantic." *French Colonial History*, 6: 17–32.

Baridon, Michel. 1985. "Ruins as a Mental Construct." *Journal of Garden History*, 5(1): 84–96.

Batey, Mavis. 1995. *Regency Gardens.* Princes Risborough, Buckinghamshire: Shire Publications Ltd.

Batsaki, Yato, et al. (eds.). 2016. "Introduction." In *The Botany of Empire in the Long Eighteenth Century*, 1–32. Washington, DC: Dumbarton Oaks Research Library and Collection.

Batsaki, Yota, Sarah Burke Cahalan, and Anatole Tchikine (eds.). 2016. *The Botany of Empire in the Long Eighteenth Century*. Washington, DC: Dumbarton Oaks.

Battie, David. 1990. *Sotheby's Concise Encyclopedia of Porcelain*. Boston and London: Little, Brown and Company.

Bauhin, Caspar. 1623. *Pinax Theatri Botanici*. Basel: Ludwig König.

Baxandall, Michael. 1985. *Patterns of Intention: On the Historical Explanation of Pictures*, New Haven and London: Yale University Press.

Baytop, Turhan. 1993. "The Tulip in Istanbul during the Ottoman Period." In Michiel Roding and Hans Theunissen (eds.), *The Tulip: A Symbol of Two Nations*, 50–6. Utrech and Istanbul: M. Th. Houtsma Stichting.

Beaglehole, J.C. (ed.). 1969. *The Voyage of the Endeavour, 1768–1771*. Cambridge: Cambridge University Press.

Beauman, Fran. 2005. *The Pineapple: King of Fruits*. London: Chatto & Windus.

Beck, Lily Y. (trans.). 2005. *Pedanius Dioscorides of Anazarbus: De Materia Medica*. Hildesheim, Germany: Olms-Weidmann.

Bending, Stephen. 2013a. *A Cultural History of Gardens in the Age of Enlightenment*. London: Bloomsbury Academic.

Bending, Stephen. 2013b. *Green Retreats: Women, Gardens and Eighteenth-Century Culture*. Cambridge: Cambridge University Press.

Benharrech, Sarah. 2018. "Botanical Palimpsests, Or Erasure of Women in Science: The Case Study of Mme Dugage du Pommereul (1733–82)." *Harvard Papers in Botany*, 23(1): 89–108.

Benkert, Davina. 2016. "The 'Hortus Siccus' as a Focal Point: Knowledge, Environment, and Image in Felix Platter's and Caspar Bauhin's Herbaria." In Susanna Burghartz et al. (eds.), *Sites of Mediation: Connected Histories of Places, Processes and Objects in Europe and Beyond, 1450–1650*, 211–39. Leiden: Brill.

Bennett, John. 1789. *Letters to a Young Lady, on a Variety of Useful and Interesting Subjects: Calculated to Improve the Heart, to Form the Manners, and Enlighten the Understanding*, 2 vols. Warrington: The Author.

Bentham, Jeremy. 2002. "Auto-Icon; Or, Farther Uses of the Dead to the Living." In James E. Crimmins (ed.), *Bentham's Auto-Icon and Related Writings*. Bristol: Thoemmes Press.

Berlin, Brent. 1999. "How a Folkbotanical System Can Be Both Natural and Comprehensive: One Maya Indian's View of the Plant World." In Douglas L. Medin and Scott Atran (eds.), *Folkbiology*, 71–89. Cambridge, MA: MIT Press.

Bernard, Thomas. n.d. [*c.* 1795]. "On the Use of Rice." A/FH/A6/20/2/1-3. Foundling Hospital Archives, London Metropolitan Archives, London.

Betts, Edwin Morris (ed.). 1944. *Thomas Jefferson's Garden Book 1766–1824, With Relevant Extracts from his Other Writings*. Philadelphia: American Philosophical Society.

Bewell, Alan. 1996. "'On the banks of the South Sea': Botany and Sexual Controversy in the Late Eighteenth Century." In David Philip Miller and Peter Hans Reill (eds.), *Visions of Empire: Voyages, Botany, and Representations of Nature*, 173–93. Cambridge: Cambridge University Press.

Bian, He. 2014. "Assembling the Cure: Materia Medica and the Culture of Healing in Late Imperial China." PhD thesis, Harvard University, USA.

Black, Jeremy. 2014. *The Power of Knowledge: How Information and Technology Made the Modern World*. New Haven: Yale University Press.

Blacker, Mary Rose. 2000. *Flora Domestica: A History of Flower Arranging 1500–1930*. London: National Trust Enterprises Ltd.

Bleecker, Ann Eliza. 1793. *The Posthumous Works of Ann Eliza Bleecker in Prose and Verse.* New York: Swords.

Bleichmar, Daniela. 2012. *Visible Empire. Botanical Expeditions and Visual Culture in the Hispanic Enlightenment.* Chicago: University of Chicago Press.

Bleichmar, Daniela. 2016. "Botanical Conquistadors. The Promises and Challenges of Imperial Botany in the Hispanic Enlightenment." In Y. Batsaki et al. (eds.), *The Botany of Empire in the Long Eighteenth Century*, 35–60. Washington, DC: Dumbarton Oaks.

Blith, Walter. 1649. *The English Improver.* London: Printed for John Wright.

Blith, Walter. 1652. *The English Improver Improved.* London: Printed for John Wright.

Boom, Brian M. 1996. "Societal and Scientific Information Needs from Plant Collections." In Tod F. Stuessy and S. H. Sohmer (eds.), *Sampling the Green World: Innovative Concepts of Collection, Preservation, and Storage of Plant Diversity*, 16–27. New York: Columbia University Press.

Boone, Chantal, and Dominique Vivent. 2013. *Trésors vivants: Valorisation des herbiers.* Bordeaux: Jardin botanique de Bordeaux.

Bougainville, Louis Antoine de. 1771. *Voyage autour du monde par la frigate du roi "la Boudeuse" et la flute "l'Étoile", en 1766, 1767, 1768 et 1769.* Paris: Saillant and Nyon.

Boumediene, Samir. 2016. *La colonisation du Savoir. Une histoire des plantes médicinales du "Nouveau Monde" (1492–1750).* Vaulx-en-Velin: Les Éditions des mondes à faire.

Bourguet, Marie-Noëlle, and Christophe Bonneuil. 1999. "De l'inventaire du monde à la mise en valeur du globe: Botanique et colonisation (Fin XVIIe siècle–début XXe siècle)." *Revue française d'histoire d'outre-mer*, 86: 9–38.

Boutroue, Marie-Elisabeth. 2008. "Pourquoi faire un herbier de plantes sèches à la Renaissance?" In Franck Lestringant (ed.), *Le Théâtre de la curiosité (XVI–XVIIe siècle)*, Actes du Colloque international du 8 mars 2007, Centre de Recherche V.-L. Saulnier, 91–107. Paris: Presse de l'université Paris-Sorbonne.

Bowerbank, Sylvia. 2004. *Speaking for Nature: Women and Ecologies of Early Modern England*, Baltimore and London: Johns Hopkins University Press.

Boym, Michal. 1656. *Flora Sinensis.* Vienna: Typis Matthaei Rictij.

Bradley, Richard. 1721. *A General Treatise of Husbandry and Gardening.* London: J. Peele.

Bravo, Michael T. 2005. "Mission Gardens: Natural History and Global Expansion, 1720–1820." In Londa Schiebinger and Claudia Swan (eds.), *Colonial Botany. Science, Commerce, and Politics in the Early Modern World*, 49–65. Philadelphia: University of Pennsylvania Press.

Breen, Benjamin. 2012. "Portugal, Early Modern Globalization and the Origins of the Global Drug Trade." *Perspectives on Europe*, 42(1): 84–8.

Bret, Patrice. 1995. "Le réseau des jardins coloniaux: Hypolite Nectoux (1759–1836) et la botanique tropicale, de la mer des Caraibes aux bords du Nil." In Yves Laissus (ed.), *Les Naturalistes français en Amérique du Sud XVIe–XIXe siècles.* Paris: Comité des Travaux Historiques et Scientifiques.

Bret, Patrice. 1999. "Des 'Indes' en Méditerranée? L'Utopie tropicale d'un jardinier des Lumières et la maîtrise agricole du territoire." *Revue français d'histoire d'outre-mer*, 86: 65–89.

Bridson, Diane, and Leonard Forman. 1992. *The Herbarium Handbook*, rev. edn. Kew: Royal Botanic Gardens.

Briggs, Robin. 1991. "The Académie Royale des Sciences and the Pursuit of Utility." *Past and Present*, 133(1): 38–88.

Brixius, Dorit. 2015. "A Pepper Acquiring Nutmeg: Pierre Poivre, The French Spice Quest and the Role of Mediators in Southeast Asia, 1740s to 1770s." *Journal of the Western Society for French History*, 43: 68–77.

Brixius, Dorit. 2017. "French Empire on the Ground: Plants, Peoples, and Knowledge in the Service of Eighteenth-Century Isle de France." PhD thesis, European University Institute, Italy.

Brockliss, Laurence. 2002. *Calvet's Web. Enlightenment and the Republic of Letters in Eighteenth-Century France*. Oxford: Oxford University Press.

Brockway, Lucile H. 1979. *Science and Colonial Expansion: The Role of the British Royal Botanic Gardens*. London and New York: Academic Press.

Brown, David, and Tom Williamson. 2016. *Lancelot Brown and the Capability Men: Landscape Revolution in Eighteenth-Century England*. London: Reaktion Books.

Browne, Janet. 1996. "Botany in the Boudoir and Garden: The Banksian Context." In David Philip Miller and Peter Hans Reill (eds.), *Visions of Empire: Voyages, Botany, and Representations of Nature*, 153–72. Cambridge: University of Cambridge Press.

Bryson, Norman. 1997. *Looking at the Overlooked. Four Essays on Still Life Painting*. Chicago: University of Chicago Press.

Bryson, Norman. 1989. "Chardin and the Text of Still Life." *Critical Inquiry*, 15(2): 227–52.

Buckler, Edward. 1640. *A Buckler Against the Fear of Death*. London: Printed by Roger Daniel.

Bulliard, Jean Baptiste François Pierre. 1783. *Dictionnaire élémentaire de la botanique ou Exposition par ordre alphabétique, des préceptes de la botanique, & de tous les termes, tant françois que latins, consacrés à l'étude de cette science*. Paris: Chez l'auteur.

Burke, Edmund. 1984. *Selected Letters of Edmund Burke*. Chicago: The University of Chicago Press.

Burrows, Simon. 2010. *A King's Ransom: The Life of Charles Théveneau de Morande, Blackmailer, Scandalmonger & Master-Spy*. London: Bloomsbury.

Burton, Audrey. 1997. *The Bukharans: A Dynastic, Diplomatic, and Commercial History, 1550–1702*. New York: Macmillan.

Cabo Gonzalez, Ana Maria. 1997. "Ibn al-Baytār et ses apports à la botanique et à la pharmacologie dans le Kitāb al-Ǧāmi." *Medièvales*, 33: 23–39.

Cain, A. J. 1997. "Locke on Species." *Archives of Natural History*, 24: 337–60.

Campbell Orr, Clarissa. 2019. *Mrs. Delany: A Life*. New Haven and London: Yale University Press.

Cantor, Paul A. 1985. "The Metaphysics of Botany: Rousseau and the New Criticism of Plants." *Southwest Review*, 70(3): 362–80.

Carney, Judith A. 2001. *Black Rice: The African Origins of Rice Cultivation*. Cambridge, MA: Harvard University Press.

Carter, Elizabeth. 1817. *Letters from Mrs. Elizabeth Carter to Mrs. Montagu between the Years 1755 and 1800*, 4 vols. London: Rivington.

Carter, H. B. 1988. *Sir Joseph Banks, 1743–1820*. London: British Museum (Natural History).

Casid, Jill H. 2005. *Sowing Empire: Landscape and Colonization*. Minneapolis and London: University of Minnesota Press.

"Catalogue/Catalogus." 1988. *The Journal of Garden History*, 8(2–3): 105–323. DOI: 10.1080/01445170.1988.10413218.

Cha-Tsu Siu, Victoria M. 2013. *Gardens of a Chinese Emperor: Imperial Creations of the Qianlong Era, 1736–1796*. Bethlehem, PA: Lehigh University Press; Lanham, MD: Rowman and Littlefield.

Cailleux, André. 1953. "La Progression du nombre d'espèces de plantes décrites de 1500 à nos jours." *Revue d'histoire des sciences*, 6(1): 42–9.

Chakrabarti, Pratik. 2005. "'Neither of meate nor drinke, but what the Doctor alloweth': Medicine Amidst War and Commerce in Eighteenth-Century Madras." *Bulletin of the History of Medicine*, 80(1): 1–38.

Chakrabarti, Pratik. 2010. "Empire and Alternatives: Swietenia febrifuga and the Cinchona Substitutes." *Medical History*, 54(1): 75–94.

Chakrabarti, Pratik. 2014. *Medicine and Empire 1600–1960*. Basingstoke: Palgrave Macmillan.

Chambers, Douglas. 1993. *The Planters of the English Landscape Garden: Botany, Trees, and The Georgics*. New Haven and London: Yale University Press.

Chartres, John. 1990. *Agricultural Markets and Trade, 1500–1750*. Cambridge: Cambridge University Press.

Chartres, John, and David Hey (eds.). 1990. *English Rural Society, 1500–1800: Essays in Honour of Joan Thirsk*. Cambridge: Cambridge University Press.

Chiem, Kristen. 2018. "Possessing the King of Flowers, and Other Things at the Qing Court." *Word & Image*, 34(4): 388–406.

Choffé, Laurent. 2004. "Le Jardin champêtre de Trianon: l'alliance du pittoresque à la botanique." *Versalia. Revue de la Société des Amis de Versailles*, 7: 56–69.

Clarke, Philip A. 2008. *Aboriginal Plant Collectors: Botanists and Australian Aboriginal People in the Nineteenth Century*. Kenthurst, New South Wales: Rosenberg Publishing.

Clay, Christopher. 1990. *Rural Society: Peasants and Labourers, 1500–1750*. Cambridge: Cambridge University Press.

Coe, Sophie D., and Michael D. Coe. 1996. *The True History of Chocolate*. New York: Thames and Hudson.

Cogan, Thomas. 1636. *The Haven of Health*. London: Printed by Anne Griffin.

Coleman, Deirdre. 2005. *Romantic Colonization and British Anti-Slavery*. Cambridge: Cambridge University Press.

Collins, Minta. 2000. *Medieval Herbals: The Illustrative Traditions*. Toronto: University of Toronto Press.

Collins, Samuel. 1664. "Report on Coffee." Russian State Archive for Ancient Documents, Moscow [RGADA] f. 143 (Apothecary Chancery), op. 2, ed. khr. 734.

Conklin, Harold C. 1980. *Ethnographic Atlas of Ifugao: A Study of Environment, Culture, and Society in Northern Luzon*. New Haven: Yale University Press.

Kelly, Christopher, and Alexandra Cook. 2000. Editors' Introduction. *Jean-Jacques Rousseau's Botanical Writings*. Hanover, NH: University Press of New England.

Cook, Alexandra. 2001. "Politics of Nature and Voyages of Exploration: Some Purposes and Results." In A. Agnarsdottir (ed.), *Voyages and Exploration in the North Atlantic from the Middle Ages to the XVIIth Century*, 125–38. Reykjavik: University of Iceland Press.

Cook, Alexandra. 2010. "Linnaeus and Chinese Plants: A Test of the Linguistic Imperialism Thesis." *Notes and Records of the Royal Society*, 64(June): 121–38.

Cook, Alexandra. 2012. *Jean Jacques Rousseau and Botany: The Salutary Science*. Oxford: The Voltaire Foundation.

Cook, Harold. 2007. *Matters of Exchange: Commerce, Medicine and Science in the Dutch Golden Age*. New Haven: Yale University Press.

Cook, Harold. J., and Timothy. D Walker. 2013. "Circulation of Medicine in the Early Modern Atlantic World." *Social History of Medicine*, 26(3): 337–51.

Cooper, Alix. 2007. *Inventing the Indigenous: Local Knowledge and Natural History in Early Modern Europe*. Cambridge: Cambridge University Press.

Cooper, Alix. 2013. "Picturing Nature: Gender and the Politics of Natural Historical Description in Eighteenth-Century Gdánsk/Danzig." *Journal for Eighteenth Century Studies*, 36(4): 519–29.

Cooper, Alix. 2018. "Placing Plants on Paper: Lists, Herbaria, and Charts as Experiments with Territorial Inventory at the Mid-Seventeenth Century Gotha Court." *History of Science*, 56 (3): 257–77.

Cornett, Peggy. 1995. "Jefferson's Botanical Perseverance." *Twinleaf*. Available online: www.monticello.org/house-gardens/center-for-historic-plants/twinleaf-journal-online/jefferson-s-botanical-perseverance (accessed May 23, 2021).

Cornett, Peggy. 2002. "'Botany & Friendship' A Circle of Transatlantic Plant Exchange." *Twinleaf*. Available online: www.monticello.org/house-gardens/center-for-historic-plants/twinleaf-journal-online/botany-and-friendship (accessed May 23, 2021).

Coşgel, Metin, M. Miceli, J. Thomas, and Jared Rubin. 2012. "The Political Economy of Mass Printing: Legitimacy and Technological Change in the Ottoman Empire." *Journal of Comparative Economics*, 40(3): 357–71.

Cowan, Brian. 2008. *The Social Life of Coffee: The Emergence of the British Coffeehouse*. New Haven and London: Yale University Press.

Cowell, John. 1651. *The Institutes of the Lawes of England*. London: Tho. Roycroft.

Cowper, William. 1782. *Poems by William Cowper*. London: J. Johnson.

Cowper, William. 1785. *The Task, A Poem in Six Books*. London: J. Johnson

Cowper, William. 1808. "Yardley Oak." *The Port Folio*, 5–6: 166–8.

Craughwell, Thomas J. 2012. *Thomas Jefferson's Creme Brulee: How a Founding Father and His Slave James Hemings Introduced French Cuisine to America*. Philadelphia: Quirk Books.

Crawford, Matthew J. 2016. *The Andean Wonder Drug: Cinchona Bark and Imperial Science in the Spanish Atlantic, 1630–1800*. Pittsburgh: University of Pittsburgh Press.

Crombie, A.C. 1994. *Styles in Scientific Thinking*, vol. 2. London: Gerald Duckworth & Co., Ltd.

Crosby, Alfred W. 2015. *Ecological Imperialism, The Biological Expansion of Europe, 900–1900*, 2nd edn. Cambridge: Cambridge University Press.

Culpeper, Nicholas. 1652. *English Physitian*. London: Printed by Peter Cole.

Culpeper, Nicholas. 1814. *Culpeper's Complete Herbal*. London: R. Evans.

Dale, Stephen F. 2002. *Indian Merchants and Eurasian Trade, 1600–1750*. Cambridge: Cambridge University Press.

Damodaran, Vinita, Anna Winterbottom, and Alan Lester (eds.). 2015. *The East India Company and the Natural World*. Basingstoke: Palgrave Macmillan.

Daniels, Stephen. 1988. "The Political Iconography of Woodland in Later Georgian England." In Denis Cosgrove and Stephen Daniels (eds.), *The Iconography of Landscape: Essays on the Symbolic Representation, Design and Use of Past Environments*, 43–82. Cambridge: Cambridge University Press.

Dannenfeldt, Karl H. 1968. *Leonhard Rauwolf: Sixteenth-Century Physician, Botanist, and Traveller*. Cambridge: Harvard University Press.

Dannenfeldt, Karl H. 1985. "Egyptian mumia: The Sixteenth Century Experience and Debate." *The Sixteenth Century Journal*, 16(2): 163–80.

Darwin, Erasmus. 1791. *Botanic Garden: A Poem in Two Parts*. London, J. Johnson.

Dauncey, Elizabeth A., et al. 2017. "Common Mistakes When Using Plant Names and How to Avoid Them." *European Journal of Integrative Medicine*, 8(5): 597–601.

Davidson, Ian. 2004. *Voltaire: A Life*. London: Profile.

de Acosta, José. 1604. *The Naturall and Morall Historie of the East and West Indies*. Translated by E.G. (attributed name Edward Grimeston). London: Printed by Val: Sims for Edward Bount and William Aspley.

de la Court, Pieter. 1737. *Byzondere aenmerkingen* .... Leiden, Abraham Kallewier.

Deerr, Noël. 1950. *History of Sugar*, 2 vols. London: Chapman and Hall.

Dekker, Cornelis. 1640. *Tractat van het excellenste kruyd thee*. N.p.

Delaporte, François. 1982. *Nature's Second Kingdom: Explorations of Vegetality in the Eighteenth Century*. Translated by A. Goldhammer. Cambridge, MA: MIT Press.

Delbourgo, James, and Nicholas Dew (eds.). 2007. *Science and Empire in the Atlantic World*. New York and Abingdon: Routledge.

Delon, Michael. 2013. *Encyclopedia of the Enlightenment*. New York: Routledge.

Demiriz, Yildiz. 1993. "Tulips in Ottoman Turkish Culture and Art." In Michiel Roding and Hans Theunissen (eds.), *The Tulip: A Symbol of Two Nations*, 57–75. Utrech and Istanbul: M. Th. Houtsma Stichting.

Diderot, Denis. 1995. *Diderot on Art, Volume I. The Salon of 1765 and Notes on Painting*. New Haven: Yale University Press.

Dietz, Bettina. 2019. "Networked Names: Synonyms in Eighteenth-Century Botany." *History and Philosophy of the Life Sciences*, 41: 46.

Dixon-Hunt, John, and Peter Willis (eds.). 1988. *The Genius of the Place: The English Landscape Garden 1620–1820*. Cambridge, MA: MIT Press.

Donne, John. 1649. *Fifty Sermons*. London: Printed by Ja. Flesher.

Drayton, Richard. 1999. "À l'école des français: La science et la deuxième empire britannique (1783–1830)." *Revue française d'histoire d'outre-mer*, 86(322–3): 91–118.

Drayton, Richard. 2000. *Nature's Government: Science, Imperial Britain and the "Improvement" of the World*. New Haven and London: Yale University Press.

Drouin, Jean-Marc. 2005. "Linné et la dénomination des vivants: Portrait du naturaliste en législateur." In Thierry Hoquet (ed.), *Fondements de la botanique: Linnaeus et la classification des plantes*, 37–55. Paris: Vuibert.

Drouin, Jean-Marc. 2008. *L'Herbier du philosophe*. Paris: Editions du Seuil.

Drouin, Jean-Marc. 2015. "Le paradoxe du Baobab. Nomenclature botanique et métissage culturel." In Philippe Selosse and Denis Reynaud (eds.), *Les mots et les choses: la langue bien faite?*, 167–74. Lyon: Presses de l'Aristoloche.

Dubos, Jean-Baptiste. 1748. *Critical Reflections on Poetry, Painting and Music. With an Inquiry into the Rise and Progress of the Theatrical Entertainments of the Ancients*, vol. 1. Translated by Thomas Nugent. London: Printed for J. Nourse.

Duhamel du Monceau, Henri Louis. 1753. *Avis pour le transport par mer des arbres, des plantes vivaces, des semences, et de diverses autres curiosités d'histoire naturelle*, 2nd edn. Paris: Imprimerie Royale.

Dupré, John. 1999. "Are Whales Fish?" In Douglas L. Medin and Scott Atran (eds.), *Folkbiology*, 461–76. Cambridge, MA: MIT Press.

Dupré, Sven, and Christoph Lüthy. 2011. *Silent Messengers: The Circulation of Material Objects of Knowledge in the Early Modern Low Countries*. Münster: LIT Verlag.

Eamon, William. 2018. "Corn, Cochineal and Quina: The Zilsel Thesis in a Colonial Iberian Setting." *Centaurus*, 60: 141–58.

Easterby-Smith, Sarah. 2016. "On Diplomacy and Botanical Gifts. France, Mysore, and Mauritius in 1788." In Y. Batsaki et al. (eds.), *The Botany of Empire in the Long Eighteenth Century*, 193–212. Washington, DC: Dumbarton Oaks.

Easterby-Smith, Sarah. 2018. *Cultivating Commerce: Cultures of Botany in Britain and France, 1760–1815*. Cambridge: Cambridge University Press.

Eliot, Jared. 1760. *Essays upon Field-Husbandry in New-England*. Boston: Edes & Gill.

Ellis, Markman. 1996. *The Politics of Sensibility: Race Gender and Commerce in the Sentimental Novel*. Cambridge: Cambridge University Press.

Ellis, Markman. 2011. *The Coffee-House: A Cultural History*. London: Weidenfeld & Nicolson.

Elman, Benjamin. 2007. "Collecting and Classifying: Ming Dynasty Compendia and Encyclopedias (Leishu)." *Extrême Orient, Extrême Occident*, hors série: 142–57.

Elvin, Mark. Forthcoming. *The Impulse of Life. The Scientific Proof of Scientific Reproduction in Plants. A Translation of Camerarius's* De sexu plantarum epistola *of 1694, with Notes, Commentaries, and Translated Background Materials*.

Ereshefsky, Marc. 2001. *The Poverty of the Linnaean Hierarchy*. Cambridge and New York: Cambridge University Press.

Erikson, E. 2015. *Chartering Capitalism: Organizing Markets, States, and Publics*. New York: Emerald Group Publishing Limited.

Evelyn, John. [1650–72] 2000. *The Diary of John Evelyn*. Edited by E. S. De Beer. Oxford: Oxford University Press.

Evelyn, John. 1776. *Sylva. Or a Discourse of Forest Trees, and the Propagation of Timber*. Annotated by Alexander Hunter. York: Ward.

Evelyn, John. 1827. *Elysium Britannicum, or The Gardens*. Edited by John E. Ingram. Philadelphia: University of Pennsylvania Press.

Evelyn, John. 2001. *Memoirs of John Evelyn ... Comprising His Diary, from 1641 to 1705-6, and a Selection of His Famliar Letters: To which is Subjoined, the Private Correspondence Between King Charles I and Sir Edward Nicholas, Also Between Sir Edward Hyde, Afterwards Earl of Clarendon, and Sir Richard Browne*. London: Henry Colburn, 1827.

Everard, William, John Palmer, John South, John Courton, William Taylor, Christopher Clifford, John Barker, Jerrard Winstanley &c. 1649. *The True Levellers Standard Advanced*. London.

Fairchild, T. 1722. *The City Gardener*. London.

Fara, Patricia. 2017. *Sex, Botany and Empire: The Story of Carl Linnaeus and Joseph Banks*. London: Icon Books.

Feingold, Richard. 1978. *Nature and Society: Later Eighteenth-Century Uses of Pastoral and Georgic*. New Brunswick: Rutger University Press.

Félibien, André. 1705. *Conférences de l'Académie royale de peinture et de sculpture*. Londres: chez David Mortier.

Finucci, Valeria. 2008. "'There's the Rub': Searching for Sexual Remedies in the New World." *Journal of Medieval and Early Modern Studies*, 38(3): 523–57.

Firth, C. H. (ed.). 1894. *The Clarke Papers, Volume II*. London: Nichols and Sons, Printed for the Camden Society.

Fludernik, Monika. 2016. "Descriptive Lists and List Descriptions." *Style*, 50(3): 309–26.

Flynn, Dennis. O., and Arturo Giráldez. 2006. "Globalization Began in 1571". In Barry K. Gills and William R. Thompson (eds.), *Globalization and Global History*, 208–22. London and New York: Routledge.

Forbes, Duncan. 1744. *Some Considerations on the Present State of Scotland*. Edinburgh: W. Sands, A. Murray, and J. Cochran.

Forster, E. S. 1927. *The Turkish Letters of Ogier Ghiselin de Busbecq, Imperial Ambassador at Constantinople, 1554–62*. Oxford: Clarendon Press.

Foucault, Michel. [1966] 1970. *The Order of Things. An Archaeology of the Human Sciences.* London and New York: Routledge.
Foundling Hospital. 1814. "No. 1. Diet Table Foundling Hospital for One Child." A/FH/A/17/017/001, Foundling Hospital Archives, London Metropolitan Archives, London.
Foust, Clifford M. 2014. *Rhubarb: The Wondrous Drug.* Princeton, NJ: Princeton University Press.
Frampton, James. 1577. *Ioyfull newes out of the newe founde worlde, wherein is declared the rare and singular vertues of diuerse and sundrie hearbes, trees, oyles, plantes, and stones, with their applications, as well for phisicke as chirurgerie.* London.
Franklin, Simon. 2010. "Printing and Social Control in Russia 1: Passports." *Russian History*, 37(3): 208–37.
Frankopan, Peter. 2015. *The Silk Roads: A New History of the World.* London: Bloomsbury Publishing.
Freedman, Paul. 2008. *Out of the East: Spices and the Medieval Imagination.* New Haven and London: Yale University Press.
Fullagar, Kate. 2010. "Reynolds' New Masterpiece: From Experiment in Savagery to Icon of the Eighteenth Century." *Cultural and Social History*, 7(2): 191–212.
Funk, Vicki. 2004. *100 Uses for the Herbarium.* New Haven: The Yale University Herbarium.
Gage, Thomas. 1648. *The English American: A New Survey of the West Indies.* London: R. Cotes.
Gänger, Stefanie. 2015. "World Trade in Medicinal Plants from Spanish America, 1717–1815." *Medical History*, 59(1): 44–62.
Garcin, Laurent. 1747. "Lettre de M. Garcin sur la nature et l'usage des pilules Madurines, avec des remarques intéressantes sur les éfets du Thé." *Journal helvétique* (May): 473–90.
Garcin, Laurent, Jacques Savary de Bruslons, and Anon. 1742. "Canelle, ou Cannelle." In J. Savary de Bruslons, *Dictionnaire universel de commerce.* Revised edn., P.-L. Savary (ed.), three vols in four parts. Vol. 1, 645–52. Geneva: Les Héritiers Cramer & Frères Philibert.
Gascoigne, John. 1998. *Science in the Service of Empire: Joseph Banks, the British State and the Uses of Science in the Age of Revolution.* Cambridge: Cambridge University Press.
Gascoigne, John. 2014. *Encountering the Pacific in the Age of Enlightenment.* Cambridge: Cambridge University Press.
George, Sam. 2005. "The Cultivation of the Female Mind: Enlightened Growth, Luxuriant Decay and Botanical Analogy in Eighteenth-Century Texts." *Journal of the History of European Ideas*, 31(2): 209–23.
Ghorbani, Abdolbaset, et al. 2018. "Botanical and Floristic Composition of the Historical Herbarium of Leonhard Rauwolf Collected in the Near East (1573–1575)." *Taxon*, 67(3): 565–80.
Giffin, Alison. 1993. "'Till *Grief* Melodious Grow': The Poems and Letters of Ann Eliza Bleecker." *Early American Literature*, 28(3): 222–41.
Gilbert, Samuel. 1682. *The Florists Vade-Mecum.* London.
Gilpin, William. 1791. *Remarks on Forest Scenery, and other Wooldand views (relative chiefly to picturesque beauty) illustrated by the scenes of New-Forest in Hampshire.* London: R. Blamire.
Glamman, Kristof. 1958. *Dutch-Asiatic Trade: 1620–1740.* Copenhagen: Springer.
Glasse, Hannah. 1747. *The Art of Cookery, Made Plain and Easy.* London: Printed for the Author.
Glaus, John, and Jean Seznec (eds.). 2011. *On Art and Artists: An Anthology of Diderot's Aesthetic Thought.* Dordrecht: Springer Netherlands.

Goethe, Johann Wolfgang von. [1790] 2009. *The Metamorphosis of Plants* 1790. Translated by Douglas Miller. Cambridge, MA: MIT Press.

Gohmann, Joanna. 2018. "Colonizing through Clay: A Case Study of the Pineapple in British Material Culture." *Eighteenth-Century Fiction*, 31(1): 143–61.

Goldgar, Anne. 1995. *Impolite Learning. Conduct and Community in the Republic of Letters, 1680–1750*. New Haven and London: Yale University Press.

Goldschmidt, Asaf. 2008. *The Evolution of Chinese Medicine: Song Dynasty, 960–1200*. London and New York: Routledge.

Gooding, Mel, David Mabberley, and Joe Studholme. 2017. *Joseph Banks' Florilegium. Botanica Treasures from the Cook's First Voyage*. London: Thames and Hudson.

Goodman, Dena. 1994. *The Republic of Letters: A Cultural History of the French Enlightenment*. Ithaca and London: Cornell University Press.

Goody, Jack. 1993. *The Culture of Flowers*. Cambridge: Cambridge University Press.

Gordon, Patrick. 2010. *Diary of General Patrick Gordon of Auchleuchries, 1635–1699: 1659–1667*. Edited by D. G. E. Fedosov. Aberdeen: AHRC Centre for Irish and Scottish Studies, University of Aberdeen.

Grant, Fiona. 2013. *Glasshouses*. London: Bloomsbury Publishing.

Greene, Edward Lee. 1983. *Landmarks of Botanical History*. Edited by Frank N. Egerton, with Robert P. McIntosh and Rogers McVaugh, part 2. Stanford: Stanford University Press.

Griffin, Clare. 2015a. "What's in a Vodka? Juniper Vodkas and the History of Distilling in Russia." Available online: http://recipes.hypotheses.org/5623 (accessed May 23, 2021).

Griffin, Clare. 2015b. "In Search of an Audience: Popular Pharmacies and the Limits of Literate Medicine in Late Seventeenth-and Early Eighteenth-Century Russia." *Bulletin of the History of Medicine*, 89(4): 705.

Griffin, Clare. 2017. "Russia and the Medical Drug Trade in the Seventeenth Century." *Social History of Medicine*, 31(1): 2–23.

Griffin, Clare. 2020. "Disentangling Commodity Histories: Pauame and Sassafras in the Early Modern Global World." *Journal of Global History*, 15(1): 1–18.

Grivetti, Louis E., and Howard-Yana Shapiro. 2009. *Chocolate: History, Culture, and Heritage*. New York: Wiley.

Gronovius, Johannes Fredericus, and Leonhard Rauwolf. 1755. *Flora orientalis, sive, Recensio plantarum quas botanicorum coryphaeus Leonardus Rauwolffus, medicus augustanus, annis 1573, 1574, et 1575, in Syria, Arabia, Mesopotamia, Babylonia, Assyria, Armenia et Judaea*. Leiden: Wilhelm de Groot.

Grove, Richard. 1996. "Indigenous Knowledge and the Significance of South-West India for Portuguese and Dutch Constructions of Tropical Nature." *Modern Asian Studies*, 30(1): 121–43.

Guédès, Michel. 1973. "Duchesne, Buisson, Durande, Early Followers of the Natural Method of the Jussieus." *Taxon*, 22(2–3): 211–19.

Guédès, Michel. 1983. "Notules de bibliographie botanique XV–XXI." *Archives of Natural History*, 11(2): 195–7.

Gunder, Frank André. 1988. *Re Orient: Global Economy in the Asian Age*. Berkeley, CA: University of California Press.

Hahn, Roger. 1971. *The Anatomy of a Scientific Institution: The Paris Academy of Sciences, 1666–1803*. Berkeley and London: University of California Press.

Hales, Stephen. 1727. "Vegetable Staticks; Or An Account of some Statical Experiments on the Sap in Vegetables." *Statical Essays*, vol. I. London: W. and J. Innys.

Hall, Marcus. 2003. “Editorial: The Native, Naturalized and Exotic—Plants and Animals in Human History.” *Landscape Research*, 28(1): 5–9.

Haller, Albrecht von. 1768. *Historia stirpium indigenarum Helvetiae inchoata*. Bern: Societatis typographicae.

Hammond, J. L., and Barbara B. Hammond. 1911. *The Village Labourer 1760–1832*. London: Longmans & Co.

Harris, Jessica. 2014. “Out of Africa: A Brief Guide to African Food History.” In Paul Freedman, Joyce E. Chaplin, and Ken Albala (eds.), *Food in Time and Place: The American Historical Association Companion to Food History*, 95–106. Oakland, CA: University of California Press.

Hartlib, Samuel. 1641. *A Description of the famous Kingdome of Macaria*. London: Printed for Francis Constable.

Hartlib, Samuel. 1648. *A further Discoverie of the Office of Publick Addresse for Accommodations*. London.

Hartlib, Samuel, and Cressy Dymock. 1651. *An Essay for Advancement of Husbandry-Learning*. London: Printed by Henry Hills.

Hartlib, Samuel, and Cressy Dymock. 1653. *A Discoverie for Division or Setting Out of Land*. London: Printed for Richard Wodenothe.

Hatch, Peter. 2003. “The Garden Plants of Lewis and Clark ‘Public Treasures’.” *Twinleaf*. Available online: www.monticello.org/house-gardens/center-for-historic-plants/twinleaf-journal-online/the-garden-plants-of-lewis-and-clark (accessed May 23, 2021).

Hatch, Peter. 2005. “‘The Work Is Very Heavy’. Gardener’s at Jefferson’s Monticello.” *Twinleaf*. Available online: www.monticello.org/sites/default/files/inline-pdfs/the_work_is_very_heavy.pdf (accessed May 23, 2021).

Hatch, Peter. 2007. “Jefferson’s Horticultural Neighborhood: A Rational Society of Gardeners.” *Twinleaf*. Available online: www.monticello.org/sites/default/files/inline-pdfs/rationalsociety.pdf (accessed May 23, 2021).

Hatch, Peter. 2012. *A Rich Spot of Earth: Thomas Jefferson’s Revolutionary Garden at Monticello*. New Haven: Yale University Press.

Haudrère, Philippe. 2005. *La Compagnie française des Indes au XVIIIe siècle*, 2 vols. Paris: Les Indes Savantes.

Heilmeyer, Marina. 2006. *The Language of Flowers: Symbols and Myths*. New York: Prestel Publishing.

Heng, Geraldine. 2014. “Early Globalities, and Its Questions, Objectives, and Methods: An Inquiry into the State of Theory and Critique.” *Exemplaria*, 26(2–3): 234–53.

Heniger, J. 1986. *Hendrik Adriaan Reede von Drakenstein and Hortus Malabaricus*. Rotterdam and Boston: A.A. Balkema.

Henrey, Blanche. 1975. *British Botanical and Horticultural Literature before 1800*, 3 vols. London: Oxford University Press.

Hernández, Francisco, and Nardo Antonio Recchi. 1651. *Rerum medicarum Novae Hispaniae thesaurus, seu, Plantarum animalium mineralium Mexicanorum historia*. Rome: Ex typographeio Vitalis Mascardi.

Herrick, Robert. 1648. *Hesperides: or, The Works Both Humane & Divine of Robert Herrick Esq*. London: Printed for John Williams, and Francis Eglesfield.

Hervey, James. 1747. *Reflections on a Flower-garden. In a Letter to a Lady*. London: J. & J. Rivington.

Hessayon, Ariel. 2008. "Restoring the Garden of Eden in England's Green and Pleasant Land: The Diggers and the Fruits of the Earth." *Journal for the Study of Radicalism*, 2(2): 1–25.
Hirvonen, Ari. 2014. "Voltaire's Garden." *Pólemos*, 8(2): 223–4.
Hobhouse, Penelope. 1992. *Plants in Garden History. An Illustrated History of Plants and Their Influence on Garden Styles from Ancient Egypt to the Present Day*. London: Pavilion.
Hochstrasser, Julie Berger. 2005. "The Conquest of Spice and the Dutch Colonial Imaginary: Seen and Unseen in the Visual Culture of Trade." In Londa Schienberger and Claudia Swan (eds.), *Colonial Botany: Science, Commerce, and Politics in the Early Modern World*, 169–86. Philadelphia, PA: University of Pennsylvania Press.
Hochstrasser, T. J. 2006. "Physiocracy and the Politics of Laissez-faire." In *The Cambridge History of Eighteenth-Century Political Thought*, 419–42. Cambridge: Cambridge University Press.
Hodacs, Hanna. 2011. "Linnaeans Outdoors: The Transformative Role of Studying Nature 'On the Move' and Outside." *British Journal for the History of Science*, 44(2): 183–209.
Hodacs, Hanna. 2016. *Silk and Tea in the North: Scandinavian Trade and the Market for Asian Goods in Eighteenth-Century Europe*. London: Palgrave Macmillan.
Hoquet, Thierry. 2005. "Linné dans le texte." In Thierry Hoquet (ed.), *Les fondements de la botanique: Linné et la classification des plantes*, 155–75. Paris: Vuibert.
Horace. [1 CE] 1926. *Satires. Epistles. The Art of Poetry*. Translated by H. Rushton Fairclough. Loeb Classical Library 194. Cambridge, MA: Harvard University Press.
Hu, Shiu-Ying. 2005. *Food Plants of China*. Hong Kong: Chinese University Press.
Huguet-Termes, Teresa. 2001. "New World Materia Medica in Spanish Renaissance Medicine: From Scholarly Reception to Practical Impact." *Medical history*, 45(3): 359–76.
Hurtado, Tomas. 1645. *Chocolate y Tabaco Ayuno Eclesiastico y Natural*. Madrid: Mercador de Libros.
Hyde, Elizabeth. 2005. *Cultivated Power: Flowers, Culture, and Politics in the Reign of Louis XIV*. Philadelphia: University of Pennsylvania Press.
Iliffe, Rob. 2008. "Science and Voyages of Discovery." In Roy Porter (ed.), *The Cambridge History of Science*, 4: 618–46. Cambridge: Cambridge University Press.
Impelluzo, Lucia. 2004. *Nature and Its Symbols*. Los Angeles: Getty Publications.
Ishizu, Mina, and Simona Valeriani. 2015. "Botanical Knowledge in Early Modern Japan and Europe: Transformations and Parallel Developments." In Susanne Friedrich et al. (eds.), *Transformations of Knowledge in Dutch Expansion*, 43–68. Berlin: De Gruyter.
Jacobson, Dawn. 1993. *Chinoiserie*. Phaidon: London.
Jacques, David. 2017. *Gardens of Court and Country: English Design 1630–1730*. New Haven and London: Yale University Press.
Jacquin, Nicolaus Joseph Freiherr von. 1763. *Selectarum Stirpium Americanarum Historia*. Vienna: ex officina Krausiana.
Jarvis, Charlie E., and Phillip Cribb. 2009. "Review: Linnaean Sources and Concepts of Orchids." *Annals of Botany*, 104(3): 1–12.
Jefferson, Thomas. 1800. "Summary of Public Service, [after 2 September 1800]." Founders Online, National Archives. Available online: https://founders.archives.gov/documents/Jefferson/01-32-02-0080 (accessed May 23, 2021). [Original source: *The Papers of Thomas Jefferson*, vol. 31, June 1, 1800–February 16, 1801, Barbara B. Oberg (ed.). Princeton, NJ: Princeton University Press, 2005, 122–5.]
Jefferson, Thomas. 2004. *The Papers of Thomas Jefferson*, Retirement Series, vol. 1, *4 March 1809 to 15 November 1809*, J. Jefferson Looney (ed.). Princeton, NJ: Princeton University Press.

Jones, Peter M. 2016. *Agricultural Enlightenment. Knowledge, Technology, and Nature, 1750–1840*. Oxford: Oxford University Press.
Jordanova, Ludmilla (ed.). 1986. *Languages of Nature: Critical Essays on Science and Literature*. London: Free Association Books.
Jussieu, Antoine-Laurent de. 1778. "Exposition d'un nouvel ordre des plantes adopté dans les démonstrations du jardin royal." In *Histoire de l'Académie royale des sciences. Année MDCCLXXIV. Avec les mémoires de mathématique et de physique, pour la même année, tirés des registres de cette Académie*, 175–97. Paris: de l'imprimerie royale.
Jussieu, Antoine-Laurent de. 1789. *Genera plantarum secundum ordines naturales disposita, juxta methodum in horto regio parisiensi exartam, anno MDCCLXXIV*. Paris: Herissant and Barrois.
Jussieu, Antoine-Laurent de. 1808. "Sixième notice historique sur le muséum." *Annales du muséum d'histoire naturelle* 11: 1–39.
Kant, Immanuel. [1790] 1953. *Critique of Judgment*. Translated by James C. Meredith. Oxford: Clarendon.
Kathirithamby-Wells, Jeyamalar. 2015. "Unlikely Partners: Malay-Indonesian Medicine and European Plant Science." In Vinita Damodaran, Anna Winterbottom, and Alan Lester (eds.), *The East India Company and the Natural World*, 193–218. Hampshire and New York: Palgrave Macmillan.
Kearney, Joy. 2012. "Agnes Block, A Collector of Plants and Curiosities in the Dutch Golden Age, and her Friendship with Maria Sibylla Merian, Natural History Illustrator." In Susan Bracken and Andrea M. Galdy (eds.), *Women Patrons and Collectors*, 67–82. Newcastle upon Tyne: Cambridge Scholars Publishing.
Kelly, Jason M. 2011. "Omai, Mai, and Joshua Reynolds: A Portrait of the Pacific Borderlands (v.1.0)". *History Working Papers*. Available online: https://wayback.archive-it.org/14458/20200715141843/http://www.historyworkingpapers.org/?page_id=73 (accessed May 23, 2021).
Kerridge, Eric. 1968. *The Agricultural Revolution*. New York: Augustus M. Kelley.
Kesseler, Rob, and Madeline Harley. 2004. *Pollen: The Hidden Sexuality of Flowers*. London: Papadakis.
Kettilby, Mary. 1734. *A Collection of Above Three Hundred Receipts in Cookery, Physick, and Surgery*. London: for W. Parker.
Kimball, Fiske. 1980. *The Creation of the Rococo Decorative Style*. New York: Dover Publications.
Kitson, Peter. 2013. *Forging Romantic China: Sino-British Cultural Exchange 1760–1840*. Cambridge: Cambridge University Press.
Klein, Wouter, and Toine Pieters. 2016. "The Hidden History of a Famous Drug: Tracing the Medical and Public Acculturation of Peruvian Bark in Early Modern Western Europe (*c.* 1650–1720)." *Journal of the History of Medicine and Allied Sciences*, 71(4): 400–21.
Kleinau, Elke. 2012. "Botany and the Taming of Female Passion: Rousseau and Contemporary Educational Concepts of Young Women." *Studies in Philosophy and Education*, 31: 465–76.
Kliger, Samuel. 1952. *The Goths in England: A Study in Seventeenth- and Eighteenth-Century Thought*. Cambridge, MA: Harvard University Press.
Koerner, Lisbet. 1999. *Linnaeus: Nature and Nation*. Cambridge: Harvard University Press.
Kopperman, Paul (ed.). 2012. *"Regimental Practice" by John Buchanan, M. D.* Farnham: Ashgate.

Kripke, Saul. 1971. "Identity and Necessity." In M. K. Munitz (ed.), *Identity and Individuation*, 135–64. New York: New York University Press.

Kumar, Deepak. 2015. "Botanical Explorations and the East India Company: Revisiting 'Plant Colonialism'." In Vinita Damodaran, Anna Winterbottom, and Alan Lester (eds.), *The East India Company and the Natural World*, 16–34. Basingstoke: Palgrave Macmillan.

Lack, Walter, and Victoria Ibáñez. 1997. "Recording Colour in Late Eighteenth-Century Botanical Drawings: Sydney Parkinson, Ferdinand Bauer and Thaddaus Haenke." *Curtis's Botanical Magazine*, 14(2): 87–100.

Lafitau, Joseph François. 1718. *Mémoire ... concernant la précieuse plante de Gin seng de Tartarie découverte en Canada ....* Paris: Joseph Mongé.

Laird, Mark. 1998) "Parterre, Grove, and Flower Garden: European Horticulture and Planting Design in John Evelyn's Time." In Therese O'Malley and Joachim Wolschke-Bulmahn (eds.), *John Evelyn's "Elysium Britannicum" and European Gardening*, 171–220. Washington, DC: Dumbarton Oaks Research Library and Collection.

Laird, Mark. 1999) *The Flowering of the Landscape Garden: English Pleasure Grounds, 1720–1800*. Philadelphia: University of Pennsylvania Press.

Laird, Mark. 2000. "From Bouquet to Basket." *The Magazine Antiques*, June.

Laird, Mark. 2013a. "Revisiting English Gardens, 1630–1730: The French Connection in Britannia." In Patricia Bouchenot-Déchin and Georges Farhat (eds.), *André Le Nôtre in Perspective*, 308–23. Paris: Château de Versailles and Hazan.

Laird, Mark. 2013b. "Plantings." In Sonja Dümpelmann (ed.), *A Cultural History of Gardens in the Age of Empire*, 69–90. London: Bloomsbury Academic.

Laird, Mark. 2014. "Greenhouse Technologies and Horticulture: The 1st Duchess of Beaufort's Badminton Florilegium (1703–5) and J. J. Dillenius's *Hortus Elthamensis* (1732)." In Michael G. Lee and Kenneth I. Helphand (eds.), *Technology and the Garden*, 55–77. Washington, DC: Dumbarton Oaks Research Library and Collection.

Laird, Mark. 2015. *A Natural History of English Gardening 1650–1800*. New Haven and London: Yale University Press.

Laird, Mark. 2017a. "An August Domain: Gardening and Collecting in the Princesses' Kew." In Joanna Marschner (ed.), *Enlightened Princesses: Caroline, Augusta, Charlotte, and the Shaping of the Modern World*, 133–62. New Haven and London: Yale University Press.

Laird, Mark. 2017b. "Lilac and Nightingale: A Heritage of Scent and Sound in Horace Walpole's Strawberry Hill." In Dede Fairchild-Ruggles (ed.), *Sound and Scent in the Garden*, 193–216. Washington, DC: Dumbarton Oaks Research Library and Collection.

Laird, Mark. 2019. "Humphry Repton at Woburn Abbey, Bedfordshire: Before and After the Red Book." *Garden History*, 47: Suppl. 1.

Laird, Mark, et al. 2020. "Schoch and 'Schrubs': Translations of a German Shrubbery Manual." *Garden History*, 48: 1.

Laird, Mark, and Alicia Weisberg-Roberts (eds.). 2009. *Mrs. Delany and her Circle*. New Haven and London: Yale University Press.

Lamarck, Jean Baptiste Pierre Antoine de Monet de. 1789. "Herbier (Herbarium)." In *Encyclopédie méthodique: botanique*, vol. 3, 111–16. Paris: Panckoucke.

Lamy, Gabriela, and Felice Olivesi. 2013. "Parterres and Floral Embellishments in the Royal Residences of Louis XIV." In Patricia Bouchenot-Déchin and Georges Farhat (eds.), *André Le Nôtre in Perspective*, 228–45. Paris: Éditions Hazan.

Lane, Joan. 1985. "'The Doctor Scolds Me': the Diaries and Correspondence of Patients in Eighteenth-Century England". In Roy Porter (ed.), *Patients and Practitioners: Lay*

Perceptions of Medicine In Pre-Industrial Society, 205–49. Cambridge: Cambridge University Press.
Larson, James L. 1967. "Goethe and Linnaeus." *Journal of the History of Ideas*, 28: 4.
Larson, James L. 1971. *Reason and Experience: The Representation of Natural Order in the Work of Carl von Linné*. Berkeley: University of California Press.
Latourrette, Marc Antoine Flerieu de Claret de, and Jean-Baptiste François Rozier. 1766. *Démonstrations élémentaires de botanique pour l'usage de l'école vétérinaire*, vol. 1. Lyon: Jean-Marie Bruyset.
Lei, Sean Hsiang-lin. 2014. *Neither Donkey nor Horse: Medicine in the Struggle over China's Modernity*. Chicago: University of Chicago Press.
Leibnitz, Gottfried Wilhelm. 1996. *New Essays on Human Understanding*. Edited by Peter Remnant and Jonathan Bennett. Cambridge: Cambridge University Press.
Leong, Elaine. 2008. "Making Medicines in the Early-Modern Household." *Bulletin of the History of Medicine*, 82: 145–68.
Lettsom, John Coakley. 1772. *The Natural History of The Tea-Tree, With Observations on the Medical Qualities of Tea, And Effects of Tea-Drinking*. London: Printed by J. Nichols for Edward and Charles Dilly.
Levi, Scott. 1999. "India, Russia and the Eighteenth-Century Transformation of the Central Asian Caravan Trade." *Journal of the Economic and Social History of the Orient*, 42(4): 519–48.
Levi, Scott. 2002. *The Indian Diaspora in Central Asia and its Trade, 1550–1900*. Leiden: Brill.
Lévi-Strauss, Claude. 1962. *La Pensée sauvage*. Paris: Plon.
Levin, Eve. 2004. "The Administration of Western Medicine in Seventeenth-Century Russia." In Jarmo Kotilaine and Marshall Poe (ed.), *Modernizing Muscovy. Reform and Social Change in Seventeenth Century Russia*, 353–79. London and New York: Routledge Curzon.
Levin, Eve. 2010. "Healers and Witches in Early Modern Russia." In Yelena Mazour-Matusevich and Alexandra S. Korros (ed.), *Saluting Aron Gurevich. Essays in History, Literature and Other Related Subjects*, 105–33. Leiden and Boston: Brill.
Levitsky, Dmitry. 1783. "Opisaniye portreta Eya Imperatorskogo Velichestva," Sobesednik ljubiteley rossiyskogo slova, soderzhashiy razniye sochineniya v stikhakh i v proze nekotorykh rossiyskikh pisateley, vol. 6. St. Petersburg: Imp. Ak. Nauk.
Lindley, John. 1836. *A Natural System of Botany, or, A Systematic View of The Organization, Natural Affinities, and Geographical Distribution, of The Whole Vegetable Kingdom*, 2nd edn. London: Longman.
Linnaeus, Carolus. 1745. *Öländska och Gothländska Resa*. Stockholm and Uppsala: Gottfried Kiesewetter.
Linnaeus, Carolus. [1761] 1786. *Dissertation on the sexes of plants*. Translated by James Edward Smith. Dublin: L. White.
Linnaeus, Carolus. 1771. *Mantissa Plantarum*, vol. 2. Stockholm: Laurentius Salvius.
Linnaeus, Carolus. 1938. *The "Critica botanica" of Linnaeus*. Translated by A.F. Hort. London: The Ray Society.
Linnaeus, Carolus. 1957–9. *Species Plantarum: A Facsimile of the First Edition 1753*, With introductory essays by William T. Stearn, 2 vols. London: The Ray Society.
Linnaeus, Carolus. 2003. *Philosophia Botanica*. Translated by Stephen Freer. Oxford: University Press.
Linschoten, J. Huygen van., P. Anton Tiele, and A. C. (Arthur Coke) Burnell. 1885. *The Voyage of John Huyghen van Linschoten to the East Indies: from the Old English Translation of 1598: the first book, containing his description of the East*. London: Hakluyt Society.

Locke, John. 1959. *An Essay Concerning Human Understanding*. Edited by Alexander Campbell Fraser. New York: Dover.
Locke, John. [1689] 2008. *Essay Concerning Human Understanding*, vol. II, chapters ix–xi: Faculties of the Mind. Oxford: Oxford University Press.
Locke, John. 1990. *Drafts for the* Essay Concerning Human Understanding, *and Other Philosophical Writings*. Edited by Peter H. Nidditch and G. A. J. Rogers. Oxford: Oxford University Press.
Locke, John. 1997. *An Essay Concerning Human Understanding*. Edited by R. Woolhouse. London: Penguin.
Lorch, Jacob. 1978. "The Discovery of Nectar and Nectaries and its Relation to Views on Flowers and Insects." *Isis*, 69(4): 514–33.
Losty, J. P., and Malini Roy. 2013. *Mughal India. Art, Culture and Empire*. London: British Library.
Loudon, John. 1822. *Encyclopedia of Gardening*. London: Longman, Hurst, Rees.
Lovejoy, Arthur O. 1932. "The First Gothic Revival and the Return to Nature." *Modern Language Notes*, 27: 7.
Low, Anthony. 1985. *The Georgic Revolution*. Princeton, NJ: Princeton University Press
Ly-Tio-Fane, Madeleine. 1970. *Mauritius and the Spice Trade: The Triumph of Jean-Nicolas Céré and his Isle Bourbon Collaborators*. Paris-La Haye: Mouton.
M. B. 1654. *The Ladies Cabinet Enlarged and Opened*. London: T.M. for M.M.G. Bedell and T. Collins.
Macfarlane, A. 1978. *The Origins of English Individualism*. Oxford: Blackwell.
Mackay, David. 1996. "Agents of Empire." In David Philip Miller and Peter Hans Reill (eds.), *Visions of Empire: Voyages, Botany, and Representations of Nature*, 38–56. Cambridge: Cambridge University Press.
Magnin-Gonze, Joëlle. 2009. *Histoire de la botanique*, rev. edn. Paris: Delachaux et Niestlé.
Mahr, August C. 1955. "Analysis of Eighteenth-Century Delaware Indian Names for Medicinal Plants." *Ethnohistory*, 2: 11–28.
Mamonov, N. E. 1881. *Materialy dlia istorii medistiny v Rossii*, 4 vols. St. Petersburg: M.M. Stasiulevich.
Manilal, Kattungal Subramaniam. 2003. *Van Rheede's Hortus Malabaricus*, English edition with annotations and modern botanical nomenclature, 12 vols. Thiruvanathapuram: University of Kerala.
Manilal, Kattungal Subramaniam, et al. 2003. "Carl Linnaeus and Hortus Malabaricus: A 250th Anniversary Tribute to Species Plantarum." *Rheedea*, 13: 3–18.
Marcon, Frederico. 2015. *The Knowledge of Nature and the Nature of Knowledge in Early Modern Japan*. Chicago: University of Chicago Press.
Marder, Michael. 2014. *The Philosopher's Plant. An Intellectual Herbarium*. New York: Columbia University Press.
Markham, Gervase. 1615a. *Countrey Contentments, in Two Bookes*. London: Printed by J.B.
Markham, Gervase. 1615b. *The English Huswife*. London: Printed by J.B.
Markham, Gervase. [1615] 1986. *The English Housewife*. Edited by Mark Best. Kingston, Ontario: McGill-Queen's University Press.
Martin, Meredith. 2011. *Dairy Queens: The Politics of Pastoral Architecture*. Cambridge, MA: Harvard University Press.
Martin, Peter. 2001. *The Pleasure Gardens of Virginia: From Jamestown to Jefferson*. Charlottesville: University of Virginia Press.

Martin, Robert. 1832. *The past and present state of the Ter trade of England ....* London: Nabu Press.

Mason, William. 1783. *The English Garden: A Poem. In Four Books.* York: Dodsley et al.

May, Robert. [1660] 1994. *The Accomplisht Cook, or, The Art and Mystery of Cookery (facsimile of 1685 edition).* Totnes, Devon: Prospect Books.

McAleer, John. 2016. "'A Young Slip of Botany': Botanical Networks in the South Atlantic, and Britain's Maritime Worlds, *c.* 1790–1810." *Journal of Global History*, 11: 24–43.

McClellan III, James E. 2010. *Colonialism and Science: Saint Domingue in the Old Regime.* Chicago: University of Chicago Press.

McClellan III, James E., and François Regourd. 2001. "The Colonial Machine: French Science and Colonization in the Ancien Régime." *Osiris*, 15: 31–50.

McClellan III, James E., and François Regourd. 2011. *The Colonial Machine: French Science and Overseas Expansion in the Old Regime.* Turnhout: Brepolis.

McNeil, Maureen. 1986. "The Scientific Muse: The Poetry of Erasmus Darwin." In L. J. Jordanova (ed.), *Languages of Nature: Critical Essays on Science and Literature*, 164–203. London: Free Association Books.

McPhail, Ian. 1979. "Linnaeus in Lapland Dress: A Puzzling Portrait." *Moreton Arboretum Quaterly*, 14(4): 49–59.

McRae, Andrew. 1996. *God Speed the Plough: The Representation of Agrarian England, 1500–1660.* Cambridge: Cambridge University Press.

Meeker, Natania, and Antónia Szabari. 2016. "Inhabiting Flower Worlds: The Botanical Art of Madeleine Françoise Basseporte." *Arts et Savoirs*, 6: 1–15.

Meiss, Millard. 1966. "Sleep in Venice. Ancient Myths and Renaissance Proclivities." *Proceedings of the American Philosophical Society*, 110(5): 348–82.

Mérigot, Madame. 1794. *La Cuisinière républicaine.* Paris: Mérigot jeune.

Métailié, Georges, and Elizabeth Hsu. 2001. "The *Bencao Gangmu* of Li Shizhen: An Innovation in Natural History?" In Elizabeth Hsu (ed.), *Innovation in Chinese Medicine*, 221–61. Cambridge: Cambridge University Press.

Métailié, Georges. 2007. "Grafting as an Agricultural and Cultural Practice in Ancient China." In Michel Conan and W. John Kress (eds.), *Botanical Progress, Horticultural Innovation and Cultural Changes.* Washington, DC: Dumbarton Oaks Research Library and Collection.

Meyers, Amy R. W., and Lisa L. Ford (eds.). 2011. *Knowing Nature: Art and Science in Philadelphia, 1740–1840.* New Haven and London: Yale University Press.

Milam, Jennifer. 2011. *Historical Dictionary of Rococo Art.* Lanham; Toronto; Plymouth, UK: Scarecrow Press.

Milam, Jennifer. 2012. "Toying with China: Cosmopolitanism and Chinoiserie in Russian Garden Design and Building Projects under Catherine the Great." *Eighteenth-Century Fiction*, 25: 115–38.

Moffett, Thomas, and Christopher Bennet. 1655. *Health's Improvement.* London: Tho. Newcomb.

Moffett, Thomas, Christopher Bennet, Robert James M.D., and William Oldys. 1746. *Health's Improvement.* London: T. Osborne.

Monahan, Erika. 2011. "Regulating Vice and Virtue: Controlling Commodities in Early Modern Siberia." In Matthews Romaniello and Tricia Starks (eds.), *Tobacco in Russian History and Culture. From the Seventeenth Century to the Present*, 61–82. London and New York: Routledge.

Monahan, Erika. 2013. "Locating Rhubarb: Early Modernity's Relevant Obscurity." In Paula Findlen (ed.), *Early Modern Things: Objects and Their Histories, 1500–1800*, 227–51. London and New York: Routledge.

Monahan, Erika. 2016. *The Merchants of Siberia: Trade in Early Modern Eurasia*. Ithaca and New York: Cornell University Press.

Montagu, Mary Wortley. 1993. *Turkish Embassy Letters*. Edited by Malcolm Jack. London: William Pickering.

Moore, Lisa. 2005. "Queer Garden: Mary Delany's Flowers and Friendships." *Eighteenth-Century Studies*, 39(1): 49–70.

Morton, A.G. 1983. "Marginalia to Andrea Cesalpino's Work on Botany." *Archives of Natural History*, 10(1): 31–6.

Mukerji, Chandra. 2005. "Dominion, Demonstration, and Domination. Religious Doctrine, Territorial Politics, and French Plant Collection." In Londa Schiebinger and Claudia Swan (eds.), *Colonial Botany: Science, Commerce, and Politics in the Early Modern World*, 19–33. Pittsburgh: University of Pennsylvania Press.

Müller-Wille, Staffan. 2007. "Collection and Collation: Theory and Practice of Linnaean Botany." *Studies in History and Philosophy of Biological and Biomedical Sciences*, 38: 541–62.

Müller-Wille, Staffan. 2017. "Names and Numbers: 'Data' in Classical Natural History 1758–1859." *Osiris*, 32: 109–28.

Murphy, Kathleen. 2013. "Collecting Slave Traders: James Petiver, Natural History, and the British Slave Trade." *William and Mary Quarterly*, 70(4): 637–70.

Myers, Amy, and Lisa Ford (eds.). 2011. *Knowing Nature: Art and Science in Philadelphia*, 1740–1840. New Haven, Yale University Press.

Nappi, Carla. 2009a. "Bolatu's Pharmacy Theriac in Early Modern China." *Early Science and Medicine*, 14(6): 737–64.

Nappi, Carla. 2009b. *The Monkey and the Inkpot: Natural History and Its Transformations In Early Modern China*. Cambridge: Harvard University Press.

Needham, Joseph, Lu Gwei-Djen, and Huang Hsin-Tsung. 1986. *Science and Civilisation in China*, vol. 6: *Biology and Biological Technology*, part I: *Botany*. Cambridge: Cambridge University Press.

Nesbitt, Mark. 2014. "Use of Herbarium Specimens in Ethnobotany." In Jan Salick, Katie Konchar, and Mark Nesbitt (eds.), *Curating Biocultural Collections: A Handbook*, 313–28. Kew: Royal Botanic Gardens.

Ness, Immanuel, and Zak Cope (eds.). 2016. *The Palgrave Encyclopedia of Imperialism and Anti-Imperialism*. New York, Palgrave McMillan.

Neumann, Joseph N. 2010. "Malabarischer Medicus—eine ethnomedizinische Quelle des frühen 18. Jahrhunderts." In Heike Liebau et al., *Mission und Forschung: Translokale Wissensproduktion zwischen Indien und Europa im 18. und 19. Jahrhundert*, 195–203. Halle: Verlag der Frankeschen Stiftung.

Newman, Jacqueline M. 2008. "Qian-Long: Qing Emperor and His Foods." *Food in History*, 15(1): 14, 17, 25 and 33. Available online: http://www.flavorandfortune.com/ffdataaccess/article.php?ID=646 (accessed May 23, 2021).

Newton, Isaac. [1704] 1979. *Opticks, Or, A Treatise of the Reflections, Refractions, Inflections & Colours of Light*. NY: Dover Publications.

Nickelsen, Kärin. 2006. *Draughtsmen, Botanists and Nature: The Construction of Eighteenth-Century Botanical Illustrations*. Dordrecht: Springer.

Nierstrasz, C. 2015. "The Popularization of Tea: East India Companies, Private Traders, Smugglers and the Consumption of Tea in Western Europe, 1700–1760." In M. Berg, F. Gottmann, H. Hodacs, and C. Nierstrasz (eds.), *Goods from the East, 1600–1800. Europe's Asian Centuries*, 263–76. London: Palgrave Macmillan.

Norton, Marcy. 2006. "Tasting Empire: Chocolate and the European Internalization of Mesoamerican Aesthetics." *The American Historical Review*, 111(3): 660–91.

O'Brien, Karen. 2009. *Women and Enlightenment in Eighteenth-Century Britain*. Cambridge: Cambridge University Press.

O'Connor, Kaori. 2013. *Pineapple. A Global History*. London: Reaktion Books.

Parker, Charles H. 2010. *Global Interactions in the Early Modern Age, 1400–1800*. Cambridge: Cambridge University Press.

Parker, Emily. 2018. "'The Taste of the Ancients': The Garden at Marble Hill." *Garden History*, 46:2.

Parkinson, John. 1629. *Paradisi in Sole Paradisus Terrestris. A Garden of all Sorts ....* London: Humfrey Lownes and Robert Young.

Parkinson, John. 1640. *Theatrum Botanicum: The Theater of Plants*. London: Tho. Cotes.

Parmentier, Jan. 1996. *Tea Time in Flanders*. Ghent: Ted Alkins Publishing.

Parsons, Christopher M. 2017. *Cultivating a New France: Knowledge, Empire and Environment in French North America, 1600–1760*. Philadelphia: University of Pennsylvania Wishnitzer.

Parsons, Christopher M., and Kathleen S. Murphy. 2012. "Ecosystems Under Sail: Specimen Transport in the Eighteenth-Century French and British Atlantics." *Early American Studies: An Interdisciplinary Journal*, 10(3): 503–39.

Paskvan, Raymond. 1971. "The Jardin du Roi: The Growth of its Plant Collections 1715–1750." PhD thesis, University of Minnesota, USA.

Payne, Christiana. 2017. *Silent Witnesses: Trees in British Art: 1760–1870*. Bristol: Sansom and Company.

Penkethman, John. 1638. *Artachthos; or, a New Booke Declaring the Assise or Weight of Bread*. London: E.G. for R.B.

Penkethman, John, and Richard Whitaker. 1648. *Accompts of Merchandise Ready Computed*. London.

Perry, Charles. 2014. "Middle Eastern Food History." In Paul Freedman, Joyce E. Chaplin, and Ken Albala (eds.), *Food in Time and Place: The American Historical Association Companion to Food History*, 107–19. Oakland, CA: University of California Press.

Pieterse, Jan N. 2012. "Periodizing Globalization: Histories of Globalization." *New Global Studies*, 6(2): 1–25.

Plotkin, Mark and Lisa Famolare (eds.). 1992. *Sustainable Harvest and Marketing of Rain Forest Products*. Washington, DC: Island Press.

Pluche, Noel Antoine. 1732–42. *Spectacle de la nature, ou Entretiens sur les particularités de l'histoire naturelle*. 9 vols. Paris.

Pluchon, Pierre. 1991. *Histoire de la colonisation française*, vol. 1. Paris: Fayard.

Pomet, Pierre, Nicolas Lémery, and Joseph Pitton. 1748. *A Complete History of Drugs. Written in French by Monsieur* Pomet, *Chief Druggist to the Late French King Lewis 14. to which is Added what is Farther Observable on the Same Subject ... Done Into English from the Originals*. London: J. and J. Bonwicke, S. Birt, W. Parker, C. Hitch, and Wicksteed.

Poor Law Commissioners. 1836. *Second Annual Report of the Poor Law Commissioners for England and Wales with Appendices A. B. C. D.* London: Charles Knight.

Pope, Alexander. 2006. *The Major Works*. Edited by Pat Rogers. Oxford: Oxford University Press.

Porter, Roy (ed.). 2008. *The Cambridge History of Science*, vol. 4, "Eighteenth-Century Science." Cambridge: Cambridge University Press.

Porter, Roy. 1985. "The Patient's View." *Theory and Society*, 14(2): 175–98.

Powell, Owen. 2003. *Galen On the Properties of Foodstuffs (De alimentorum facultatibus)*. Cambridge: Cambridge University Press.

Pratt, Mary Louise. 1992. *Imperial Eyes. Travel Writing and Transculturation*. London and New York: Routledge.

Proctor, Robert N. 1995. *The Cancer Wars: How Politics Shapes What We Know and Don't Know About Cancer*. New York: Basic Books.

Pugliano, Valentina. 2013. "Botanical Artisans: Apothecaries And The Study Of Nature In Venice And London, 1550–1610." PhD thesis, Oxford, UK.

Quinn, Frederick. 2000. *The French Overseas Empire*. Westport and London: Praeger.

Raj, Kapil. 2007. *Relocating Modern Science: Circulation and the Construction of Knowledge in South Asia and Europe, 1650–1900*. Basingstoke: Palgrave Macmillian.

Raj, Kapil. 2016. "Rescuing Science from Civilization: On Joseph Needham's Asiatic Mode of (Knowledge) Production." In Arun Bala and Prasenjit Duara (eds.), *The Bright Dark Ages*, 255–80. Leiden: Brill.

Raj, Kapil. 2017. "Thinking Science without the Scientific Revolution: Global Interactions and the Construction of Knowledge." *Journal of Early-Modern Science*, 21: 445–58.

Rauwolf, Leonhard. 1582. *Aigentliche Beschreibung der Raiß inn die Morgenländer*. Laugingen: Leonhart Reinmichel.

Raven, Charles E. 1947. *English Naturalists from Neckam to Ray: A Study of the Making of the Modern World*. Cambridge: Cambridge University Press.

Raven, Charles E. 1950. *John Ray, Naturalist: His Life and Works*. Cambridge: Cambridge University Press.

Ray, John. 1682. *Methodus Plantarum Nova*. London: Henry Faitborne & John Kersey.

Reede tot Drakenstein, Hendrik Adriaan van. 1678–1703. *Hortus Indicus Malabaricus*, 12 vols. Amsterdam: Johan van Someren and Jan van Dyck.

Renner, Andreas. 2010. *Russische Autokratie und Europäische Medizin: Organisierter Wissenstransfer im 18. Jahrhundert*. Wiesbaden: Franz Steiner Verlag.

Reynolds, Joshua. 1837. *Discourses on Painting and the Fine Arts, delivered at the Royal Academy*. London: Jones & Company.

Riddell, John N. D. 1986. "John Parkinson's Long Acre Garden 1600–1650." *The Journal of Garden History*, 6(2): 112–24.

Riello, Giorgio. 2009. "The Globalisation Of Cotton Textiles: Indian Cottons, Europe and the Atlantic World, 1600–1850." In Giorgio Riello and Prasannan Parthasarathi (eds.), *The Spinning World: A Global History of Cotton Textiles, 1200–1850*, 261–87. Oxford: Oxford University Press.

Roberts, Lissa. 2014. "'*Le centre de toutes choses*': Constructing and Managing Centralization on the Isle de France." *History of Science*, 52(3): 319–42.

Robinson, Benedict S. 2009. "Green Seraglios: Tulips, Turbans, and the Global Market." *Journal for Early Modern Cultural Studies*, 9(1): 92–122.

Roger, Jacques. 1997. *Buffon: A Life in Natural History*. Translated by Sarah L. Bonnefoi; edited by L. Pearce Williams. Ithaca: Cornell University Press.

Roland Michel, Marianne. 1996. *Chardin*. New York: Harry N. Abrams.

Romaniello, Matthew P. 2016. "True Rhubarb? Trading Eurasian Botanical and Medical Knowledge in the Eighteenth Century." *Journal of Global History*, 11(1): 3–23.

Ross, Ian Campbell. 1981. "'Every One to Cultivate his Own Garden': John Evelyn, Voltaire and Candide." *Notes & Questions*, 28(3): 234–7.

Rousseau, Jacques. 1970. "Sébastien Vaillant: An Outstanding 18th Century Botanist." In *Essays in Biohistory and other Contributions presented by Friends and Colleagues to Frans Verdoorn*, P. Smit and R.J.Ch.V. ter Laage (eds.), 195–228. Utrecht: International Association for Plant Taxonomy.

Rousseau, Jean-Jacques. 1762. *Emile, or on education*, vol. 1. Francfort: n.p.

Rousseau, Jean-Jacques. 2000. "Botanical Writings." In A. Cook (trans. and ed.), C. Kelly (vol. ed.), *Collected Writings of Rousseau*, vol. 8. Hanover: University Press of New England.

Rowell, Margery. 1978. "Russian Medical Botany Before the Time of Peter the Great." *Sudhoffs Archiv*, 62(4): 339–58.

Rowell, Margery. 1980. "Linnaeus and Botanists in Eighteenth-Century Russia." *Taxon*, 29(1): 15–26.

Rubel, William. 2011. *Bread: A Global History*. London: Reaktion Books.

Rumphius, Georg Everhard. 1741–50. *Herbarium amboinense*, 6 vols. Edited by J. Burman. Amsterdam: Meinard Uytwerf.

Rundell, Maria Eliza. 1824. *A New System of Domestic Cookery*. London: John Murray.

Sachs, Julius. 1875. *Geschichte der Botanik vom 16. Jahrhundert bis 1860*. Munich: R. Oldenbourg.

Safier, Neil. 2008. "Fruitless Botany. Joseph de Jussieu's South American Odyssey." In James Delbourgo and Nicholas Dew (eds.), *Science and Empire in the Atlantic World*, 203–24. New York and London: Routledge.

Sainsbury, Ethel Bruce. 1925. *A Calendar of the Court Minutes Etc. of the East Iindia Company 1664–1667*. Oxford: Clarendon Press.

Saint-Lager, Jean-Baptiste. 1885. *Histoire des herbiers*. Paris: J.B. Ballière.

Saint-Simon, Louis de Rouvroy, duc de. 2006. *The Memoirs of Louis XIV*. Project Gutenberg eBook.

Salzmann, Ariel. 2000. "The Age of Tulips: Confluence and Conflict in Early Modern Consumer Culture (1550–1730)." In Donald Quataert (ed.), *Consumption Studies History of the Middle East and the History of the Ottoman Empire, 1550–1922*, 83–106. Albany, NY: State University of New York.

Samayoa, Marianne B. 2006. "More than Quacks: Seeking Medical Care in Late Colonial New Spain." *Social History of Medicine*, 19(1): 1–18.

Samper, Mario, and Radin Fernando. 2003. "Historical Statistics of Coffee Production and Trade from 1700–1960." In W. Clarence-Smith and S. Topik (eds.), *The Global Coffee Economy in Africa, Asia and Latin America, 1500–1989*. Cambridge: Cambridge University Press.

Schabas, Margaret. 2005. *The Natural Origins of Economics*. Chicago: University of Chicago Press.

Schaffer, Simon, Lissa Roberts, Kapil Raj, and James Delbourgo (eds.). 2009. *The Brokered World. Go-Betweens and Global Intelligence, 1770–1820*. Sagamore Beach: Science History Publications.

Scheid, Volker. 2013. "The Treatise on Cold Damage as a Window on Emergent Formations of Medical Practice in East Asia." *Asian Medicine*, 8(2): 295–8.

Schiebinger, Londa. 1991. "The Private Life of Plants: Sexual Politics in Carl Linnaeus and Erasmus Darwin." In Marina Benjamin (ed.), *Science and Sensibility: Gender and Scientific Enquiry, 1780–1945*, 121–43. Oxford: Blackwell.

Schiebinger, Londa. 1993. "Why Mammals are Called Mammals: Gender Politics in Eighteenth-Century Natural History." *The American Historical Review*, 98(2): 382–411.
Schiebinger, Londa. 2004a. *Nature's Body: Gender in the Making of Modern Science*. New Brunswick, NJ: Rutgers University Press.
Schiebinger, Londa. 2004b. *Plants and Empire. Colonial Bioprospecting in the Atlantic World.* Cambridge and London: Harvard University Press.
Schiebinger, Londa. 2017. *Secret Cures of Slaves: People, Plants, and Medicine in the Eighteenth-Century Atlantic World*. Stanford: Stanford University Press.
Schiebinger, Londa, and Claudia Swan (eds.). 2005. *Colonial Botany. Science, Commerce, and Politics in the Early Modern World*. Philadelphia: University of Pennsylvania Press.
Schivelbusch, Wolfgang. 1992. *Tastes of Paradise. A Social History of Spices, Stimulants and Intoxicants*. Translated by David Jacobson. New York: Pantheon.
Schlegel, G. 1900. "First Introduction of Tea into Holland." *T'oung Pao*. Second Series, 1: 468–72.
Schneider, Norbert. 1996. *The Art of the Flower*. Switzerland: Edition Stemmle.
Schweikardt, C. 2003. "More than Just a Propagandist for Tea: Religious Argument and Advice on a Healthy Life in the Work of the Dutch Physician Cornelis Bontekoe (1647–1685)." *Medical History*, 47(3): 357–68.
Scott, Christina. 2010. "The VOC and Swedish Natural History. The Transmission of Scientific Knowledge in the Eighteenth Century." In Siegfried Huigen, Elmer Kolfin, and Jan de Jong (eds.), *The Dutch Trading Companies as Knowledge Networks*, 361–92. Leiden: Brill.
Scott, Katie. 1995. *The Rococo Interior: Decoration and Scoial Spaces in Early Eighteenth-Century Paris*. New Haven and London: Yale University Press.
Segal, Sam. 1990. *Flowers and Nature*. Amstelveen: Hijnk International.
Selosse, Philippe. 2005. "The Underlying Pattern of the Renaissance Botanical Genre Pinax." In Janne Skaffari et al. (eds.), *Opening Windows on Texts and Discourses of the Past*, 161–78. Amsterdam and Philadelphia: John Benjamins.
Selosse, Philippe. 2008. "Les concepts de genre et d'espèce à travers l'évolution du modèle logique de la définition, de l'Antiquité à Linné". In Daniel Prat, Aline Raynal-Roques, Albert Roguenant, and Jean-Clude Lacharpagne (eds.), *Peut-on classer le vivant? Linné et la systématique aujourd'hui*, 65–79. Paris: Belin.
Sengupta, Jayanta. 2014. "India." In Paul Freedman, Joyce E. Chaplin, and Ken Albala (eds.), *Food in Time and Place: The American Historical Association Companion to Food History*, 68–94. Oakland, CA: University of California Press.
Seward, Anna. 1811. Letter to William Hayley, Lichfield, December 23, 1784 in *Letters of Anna Seward: Written Between the Years 1784 and 1807*, 4 vols. Edinburgh: Archibald Constable.
Shammas, Carole. 1990. *The Pre-Industrial Consumer in England and America*. New York: Oxford University Press.
Shaw, Christopher. 1989. *The Imagined Past: History and Nostalgia*. Manchester: Manchester University Press.
Shefer-Mossensohn, Miri. 2010. *Ottoman Medicine: Healing and Medical Institutions, 1500–1700*. Albany: SUNY Press.
Shenstone, William. 1764. *The Works in Verse and Prose, of William Shenstone, Esq;: Essays on men, manners, and things. A description of The Leasowes, the seat of the late William Shenstone*. London: R. and J. Dodsley, in Pall-mall.

Sheridan, Richard B. 1974. *Sugar and Slavery: An Economic History of the British West Indies, 1623–1775*. Baltimore: The Johns Hopkins University Press.

Sherry, Frank. 2009. *Raiders and Rebels: A History of the Golden Age of Piracy*. New York: Harper Collins.

Short, Thomas. 1730. *A dissertation upon tea, explaining its nature and properties, by many new experiments; and demonstrating from philosophical principles, the various effects it has on different constitutions ....* London: Printed by W. Bowyer, for F. Gyles.

Shteir, Ann. 1995. "'The Pleasing Objects of Our Present Researches': Women in Botany." In Valerie Frith (ed.), *Women & History: Voices of Early-Modern England*, 145–63. Toronto: Coach House Press.

Shteir, Ann. 1996. *Cultivating Women, Cultivating Science: Flora's Daughters and Botany in England, 1760–1860*. Baltimore and London: The Johns Hopkins University Press.

Sikkens-De Zwaan, Marisca. 2002. "Magdalena Poulle (1632–99): A Dutch Lady in a Circle of Botanical Collectors." *Garden History*, 30(2): 206–20.

Silver, Sean R. 2008. "Locke's Pineapple and the History of Taste." *The Eighteenth Century*, 49 (1): 43–65.

Simo, Melanie Louis. 1988. *Loudon and the Landscape: From Country Seat to Metropolis*. New Haven and London: Yale University Press.

Singh, Anjana. 2015. "Botanical Knowledge in Early Modern Malabar and the Netherlands: A Review of Van Reede's Hortus Malabaricus." In S. Friedrich et al., *Transformations of Knowledge in Dutch Expansion*, 187–207. Berlin: De Gruyter.

Siraisi, Nancy G. 1990. *Medieval and Early Renaissance Medicine: An Introduction to Knowledge and Practice*. Chicago: University of Chicago Press.

Siraisi, Nancy. G. 2007. *History, Medicine, and the Traditions of Renaissance Learning*. Ann Arbor, MI: University of Michigan Press.

Sivasundaram, Sujit. 2010. "Sciences and the Global: On Methods, Questions, and Theory." *Isis*, 101: 146–58.

Skelton, Robert. 1972. "A Decorative Motif in Mughal Art." In Pratapaditya Pal (ed.), *Aspects of Indian Art*. Leiden: Brill.

Slaughter, Mary M. 1982. *Universal Languages and Scientific Taxonomy in the Seventeenth Century*. Cambridge: Cambridge University Press.

Sloan, Kim (ed.). 2003. *Enlightenment: Discovering the World in the Eighteenth Century*. London: The British Museum Press.

Smart, Christopher. 1764. *Ode to the Right Honourable the Earl of Northumberland ... With some other Pieces*. London: Dodsley.

Smith, Sir James Edward. 1804–5. *Exotic Botany*, 2 vols. London: R. Taylor and Co.

Sokoloff, Dimitry D., et al. 2002. "The History of Botany in Moscow and Russia in the 18th and Early 19th Centuries in the Context of the Linnaean Collection at Moscow University (MW)." *Huntia*, 11(2): 129–91.

Spary, Emma C. 2000. *Utopia's Garden. French Natural History from Old Regime to Revolution*. Chicago and London: University of Chicago Press.

Spary, Emma C. 2004. "'Peaches which the Patriarchs Lacked': Natural History, Natural Resources, and the Natural Economy in Eighteenth-Century France." *History of Political Economy*, supplement to vol. 35. Edited by Neil De Marchi and Margaret Schabas: 14–41.

Spary, Emma C. 2005. "Of Nutmegs and Botanists: The Colonial Cultivation of Botanical Identity." In Londa Schiebinger and Claudia Swan, *Colonial Botany: Science, Commerce, and Politics in the Early Modern World*. Pittsburgh, PA: University of Pennsylvania Press.

Spary, Emma C. 2012. *Eating the Enlightenment: Food and the Sciences in Paris, 1670–1760.* Chicago: University of Chicago Press.

Speed, Adolphus. 1659. *Adam out of Eden*. London.

Spencer, Jane. 2012. "'The Link which Unites Man with Brutes': Enlightenment Feminism, Women and Animals." *Intellectual History Review*, 22(3): 427–44.

Spudich, Annamma. 2008. *Such Rich Treasure and Merchandize. Indian Botanical Knowledge in 16th and 17th Century Botanical European Botanical Books*. Bangalore: NCBS.

Stafford, Fiona. 2016. *Long, Long Life of Trees*. New Haven and London: Yale University Press.

Stearn, William T. 1957. "The *Amoenitates academicae* and the Authorship of Linnean Dissertations." In Carl Linnaeus, *Species Plantarum: A Facsimile of the First edition 1753*, vol. I. London: Ray Society.

Stearn, William T. 1959. "The Background of Linnaeus's Contributions to the Nomenclature and Methods of Systematic Biology." *Systematic Zoology*, 8: 4–21.

Stearn, William T. 1992. *Botanical Latin: History, Syntax, Terminology and Vocabulary*, 4th edn. Newton Abbot: David and Charles.

Stevens, Peter F. 1994. *The Development of Biological Systematics: Antoine-Laurent de Jussieu, Nature, and the Natural System*. New York: Columbia University Press.

Stevens, Peter F. 2002. "Why Do We Name Organisms?" *Taxon*, 51: 11–26.

Stewart, Larry. 2008. "Global Pillage: Science, Commerce and Empire." In Roy Porter (ed.), *The Cambridge History of Science*, 4: 825–44. Cambridge: Cambridge University Press.

Stewart, Margaret. 2016. *The Architectural, Landscape and Constitutional Plans of the Earl of Mar, 1700–32*. Dublin: Four Courts Press.

Strong, Roy. 1979. *The Renaissance Garden in England*. London: Thames and Hudson.

Strong, Roy. 2003. *Feast: A History of Grand Eating*. New York: Houghton Mifflin Harcourt.

Swift, Thomas. [*c*. 1720] 1801. *An Essay on Modern Education*. In John Nichols (ed.), *The Works of the Rev. Jonathan Swift*, vol. 5. London: H. Baldwin and Son.

Switzer, Stephen. 1728. *A compendious, but more particular method, than has ever yet been published, for the raising Italian brocoli, Spanish cardoon, &c*. London: Printed by S.A.

Symes, Michael. 2013. "Plantings." In Stephen Bending (ed.), *A Cultural History of Gardens in the Age of Enlightenment*, 73–92. London: Bloomsbury Academic.

Theophrastus. 1916. *Enquiry into Plants*, vol. I. Translated by Sir Arthur Hort. London and Cambridge: Harvard University Press.

Thick, Malcolm. 1990. "Roots and Other Garden Vegetables in the Diet of Londoners. *c*. 1550–1650, and Some Responses to Harvest Failure on the 1590s." In Harlan Walker (ed.), *Proceedings of the Oxford Symposium on Food & Cookery 1989: Staple Foods*, 228–36. London: Prospect Books.

Thick, Malcolm. 1998. *The Neat House Gardens: Early Market Gardening Around London*. Totnes: Prospect Books.

Thick, Malcolm. 2014. "Three 'Gentleman's Ideal Farms' and the Influence of London upon Them." Version of paper presented at The History of Gardens and Landscapes Seminar, Institute of Historical Research, Universty of London, January 31, 2012.

Thiem, Barbara, and Elisabeth Buk-Berge. 2017. "Twinflower (Linnaea borealis L.)—Plant Species of Potential Medicinal Properties." *Herba Polonica*, 63(3): 56–64.

Thirsk, Joan (ed.). 1985–9. *The Agrarian History of England and Wales*. Cambridge: Cambridge University Press.

Thirsk, Joan. 1990. *Agricultural Change: Policy and Practice, 1500–1750*, vol. 3. Cambridge: Cambridge University Press.

Thirsk, Joan. 2000. *Alternative Agriculture: A History: From the Black Death to the Present Day*. Oxford: Oxford University Press.

Thirsk, Joan. 2007. *Food in Early Modern England: Phases, Fads, Fashions 1500–1760*. London: Hambledon Continuum.

Thirsk, Joan, and J.P. Cooper (eds.). 1972. *Seventeenth-Century Economic Documents*. Oxford: Clarendon Press.

Thomas, Keith. 1983. *Man and the Natural World. Changing Attitudes in England 1500–1800*. London: Allen Lane.

Thun, Maria, and Angelika Throll-Keller. 1999. *Gardening for Life—the Biodynamic Way*. Stroud: Hawthorn Press.

Tibbles, Anthony. 1994. *Transatlantic Slavery: Against Human Dignity*. London: HMSO.

Tiele, Pieter Anton (ed.). 1885. *The voyage of John Huyghen van Linschoten to the East Indies: from the old English translation of 1598: the first book, containing his description of the East, in two volumes*, vol. II. London, Whiting & Co.

Tobin, Beth. 1996. "Imperial Designs: Botanical Illustration and the British Botanic Empire." *Studies in Eighteenth-Century Culture*, 25: 265–92.

Tomasi, Lucia. 1997. *An Oak Spring Flora: Flower Illustration from the Fifteenth Century to the Present Time: A Selection of the Rare books, Manuscripts, and Works of Art in the Collection of Rachel Lambert Mellon*. New Haven and London: Yale University Press.

Toscan, Georges. 1800. *L'Ami de la nature, ou Choix d'observations sur divers objets de la nature et de l'art*. Paris: Cuchet.

Tournefort, Joseph Pitton de. 1694. *Elemens de botanique*, vol. 1. Paris: de l'Imprimerie royale.

Tournefort, Joseph Pitton de. 1700. *Institutiones rei herbariae*. Paris: de l'Imprimerie Royale.

Trambaiolo, Daniel. 2013. "Native and Foreign in Tokugawa Medicine." *The Journal of Japanese Studies*, 39(2): 299–324.

Ukers, William H. 1922. *All About Coffee*. New York: The Tea and Coffee Trade Journal Company.

Ukers, William H. 1935. *All About Tea*. 2 vols. New York: The Tea and Coffee Trade Journal Company and Kinsport Press.

Ussher, Judith. 1871. Letter of July 27, 1796 in *Extracts from the Letters of Elizabeth, and Judith Ussher, late of the city of Waterford, Ireland*. Philadelphia: W.H. Pile's Son.

Vaillant, Sébastien. 1718. *Sermo de structura florum: Discours sur la structure des fleurs*. Leiden: Pieter van der Aa.

Van Andel, Tinde. 2015. "The Reinvention of Household Medicine by Enslaved Africans in Suriname." *Social History of Medicine*, 29(4): 676–94.

Van den Spiegel, Adriaan. 1606. *Isagoges in rem herbariam libri duo*. Padua: Paolo Meietti.

Veluwenkamp, Jan. W., and Werner Scheltjens. 2017. "Baltic Drugs Traffic, 1650–1850. Sound Toll Registers Online as a Source for the Import of Exotic Medicines in the Baltic Sea Area." *Social History of Medicine*, 31(1): 140–76.

Vogel, Virgil. J. 2013. *American Indian Medicine*. Norman: University of Oklahoma Press.

Wallis, Patrick. 2012. "Exotic Drugs and English Medicine: England's Drug Trade, *c*. 1550–*c*. 1800." *Social History of Medicine*, 25(1): 20–46.

Walpole, Horace. [1780] 1995. *The History of the Modern Taste in Gardening*. [London] New York: Ursus Press.

Walpole, Horace. 1780. "On Modern Gardening" (1770), in his *Anecdotes of Painting in England*, 4 vols. London: Dodsley.

Walvin, James. 1997. *Fruits of Empire: Exotic Produce and British Taste. 1660–1800*. Basingstoke and London: Macmillan.

Waxman, Sandra R. 1999. "The Dubbing Ceremony Revisited: Object Naming and Categorization in Infancy and Early Childhood." In Scott Atran and Douglas L. Medin (eds.), *Folkbiology*, 233–84. Cambridge, MA: MIT Press.

Way, Twigs. 2006. *Virgins, Weeders and Queens: A History of Women in the Garden*. Stroud: Sutton Publishing

Weber, Caroline. 2008. *Queen of Fashion: What Marie Antoinette Wore to the Revolution*. London: Aurum.

Webster, Charles. 2008. *Paracelsus: Medicine, Magic and Mission at the End of Time*. New Haven: Yale University Press.

Weisberg-Roberts, Alicia. 2009. "Introduction (1): Mrs Delany from Source to Subject." In Mark Laird and Alicia Weisberg-Roberts, *Mrs Delany and Her Circle*, 1–19. New Haven: Yale University Press.

Welch, Ellen R. 2017. "Picturing Europe." *Journal18*, May. Available online: http://www.journal18.org/1812 (accessed May 23, 2021).

Wells, Roger A.E. 1988. *Wretched Faces: Famine in Wartime England, 1793–1801*. Gloucester: Alan Sutton.

Weststeijn, Arthur. 2012. *Commercial Republicanism in the Dutch Golden Age: The Political Thought of Johan & Pieter de la Court*. Leiden: Brill.

Wharncliffe, Lord (ed.). 1837. Letter from Lady Mary Wortley Montagu to Mary Wortley Montagu Stuart, Countess of Bute, March 6, 1753. In *The Letters and Works of Lady Mary Wortley Montagu*, vol. 3, 239–42. London: Richard Bentley.

Wijnands, D.O. 1988. "Hortus Auriaci: The Gardens of Orange and Their Place in Late 17th-Century Botany and Horticulture." *The Journal of Garden History*, 8(2): 61–86.

Wild, Anthony. 2005. *Coffee: A Dark History*. London: W.W. Norton.

Wilkins, John S. 2011. *Species: A History of the Idea*. Berkeley: University of California Press.

Will, Pierre-Etienne, and Roy Bin Wong. 1991. *Nourish the People: The State Civilian Granary System in China, 1650–1850*. Ann Arbor, MI: Center for Chinese Studies, University of Michigan.

Williams, Raymond. 1976. *Keywords: A Vocabulary of Culture and Society*, "Nature". London: Croom Helm.

Williams, Robert. 2000. "Vanbrugh's India and his Mausolea for England." In Christopher Ridway and Robert Williams (eds.), *Sir John Vanbrugh and Landscape Architecture in Baroque England 1690–1730*. Stroud: Sutton Publishing Limited.

Williams, Roger L. 2001. *Botanophilia in Eighteenth-Century France: The Spirit of the Enlightenment*. Dordrecht and Boston: Kluwer Academic Publishers.

Wills, K.J. (ed.). 2017. *State of the World's Plants 2017: Report*. Royal Botanic Gardens Kew.

Wimmer, Clemens Alexander. 2001. *Bäume und Sträucher in historischen Gärten: Gehölzverwendung in Geschichte und Denkmalpflege*. Dresden: Verlag der Kunst.

Winsor, Mary P. 2009. "Taxonomy was the Foundation of Darwin's Evolution." *Taxon*, 58(1): 43–9.

Winstanley, Gerrard. 1649. *The New Law of Righteousness*. London: Printed for Giles Calvert.

Winstanley, Gerrard, John Barker, and Thomas Star. 1649. *An Appeal to the House of Commons*. London: Printed for Giles Calvert.

Winstanley, Jerrard. 1652. *The Law of Freedom in a Platform*. London: Printed by J.M.

Winterbottom, Anna. 2014. "Of the China Root: A Case Study of the Early Modern Circulation of Materia Medica." *Social History of Medicine*, 28(1): 22–44. DOI: 10.1093/shm/hku068.

Winterbottom, Anna. 2016. *Hybrid Knowledge in the Early East India Company World.* Basingstoke: Palgrave Macmillan.

Wishnitzer, Avner. 2014. "Into the Dark: Power, Light and Nocturnal Life in 18th-Century Istanbul." *International Journal of Middle East Studies*, 46(3): 513–31.

Woldbye, Vibeke. 1991. *Flowers into Art. Floral Motifs in European Painting and Decorative Arts*. The Hague: SDU Publishers.

Wong, Young-Tsu. 2001. *A Paradise Lost. The Imperial Garden Yuanming Yuan*. Honolulu: University of Hawaii Press.

Woolsey, Sarah Chauncey (ed.) 1879. Letter from Mary Delany to Anne Dewes [?1763], *The Autobiography and Correspondence of Mrs. Delany*, 2: 102. Boston, MA: Roberts Bros.

Woudstra, Jan. 2014. "'Much better contrived and built then [*sic*] any other in England'. Stoves and Other Structures for the Cultivation of Exotic Plants at Hampton Court Palace, 1689–1702." In Michael G. Lee and Kenneth I. Helphand (eds.), *Technology and the Garden*, 79–107. Washington, DC: Dumbarton Oaks Research Library and Collection.

Wujastyk, Dominik. 2003. *The Roots of Ayurveda: Selections from Sanskrit Medical Writings*. New Delhi: Penguin.

Yonan, Michael. 2011. *Empress Maria Theresa and the Politics of Habsburg Imperial Court*. Philadelphia: Penn State University Press.

Young, Arthur (ed.). 1796. *Annals of Agiculture, and Other Useful Arts*. Bury St. Edmunds: J. Rackham.

CONTRIBUTORS

Stephen Bending is Professor of Eighteenth-Century Literature and Culture at the University of Southampton, UK. His books include *The Writing of Rural England* (2003), *Green Retreat: Women, Gardens and Eighteenth-Century Culture* (2013), and *A Cultural History of Gardens in the Age of Enlightenment* (2013). His next book is on the problems of pleasure in pleasure gardens.

Alexandra Cook is Professor of Philosophy at the University of Hong Kong. Her books include the award-winning *Jean-Jacques Rousseau and Botany, the Salutary Science* (2012), a translation of Rousseau's botanical writings in *Collected Writings of Rousseau*, vol. 8 (2000), and *Bioprospecting Asia: Laurent Garcin and the Circulation of Scientific Knowledge* (forthcoming).

Sarah Easterby-Smith is Senior Lecturer in History at the University of St. Andrews, Scotland, UK. She is the author of the 2018 monograph *Cultivating Commerce: Cultures of Botany in Britain and France, 1760–1850*. Her current research project, "Making Knowledge, Forging Empire," examines the French collection of data about the natural world around the Indian Ocean rim in the second half of the eighteenth century.

Clare Griffin is Assistant Professor of the History of Science and Technology at Nazarbayev University, Republic of Kazakhstan. She is a specialist in science in the premodern Russian Empire and is currently completing her first book, *Material Worlds: Materia Medica and the Global History of Early Modern Russian Medicine*.

Ekaterina Heath is Research Fellow at the University of Sydney, Australia. Her recent articles include "Sowing the Seeds for Strong Relations" (2017), "Grand Tour Memories in Pavlovsk Park" (2021), and "Giving Women History" (2021). Her upcoming book focuses on Russian Empress Maria Fedorovna's use of garden spaces to influence politics.

Mark Laird is a heritage planting consultant, notably for Painshill Park Trust, Surrey, UK, and he advises on cultural landscapes in North America and Europe. He is Professor in the Landscape Architecture Program at University of Toronto, Canada. His books include *The Flowering of the Landscape Garden* (1999) and *A Natural History of English Gardening* (2015).

Jane Levi is a visiting research fellow at King's College London, Managing Editor of *Stone Soup* magazine, and co-founder of Edible Utopia, a creative urban farming project resident at Somerset House, London. Her publications include the co-authored book *Food, Politics, and Society: Social Theory and the Modern Food System* (2018).

Jennifer Milam is Pro Vice Chancellor and Professor of Art History at the University of Newcastle, Australia. Her books include *Women, Art and the Politics of Identity in Eighteenth-Century Europe* (2003), *Fragonard's Playful Paintings* (2007), *Historical Dictionary of Rococo Art* (2011), *Beyond Chinoiserie* (2018), and *Making Ideas Visible* (2022). Her next book addresses cosmopolitan ideals in garden spaces.

Garritt Van Dyk is Lecturer in History at the University of Newcastle, Australia. Wider research interests include the history of empire, early modern economic history, and European patterns of consumption in the Enlightenment. He received the Sophie Coe Prize for Writing in Food History in 2014.

INDEX